A POCKET GUIDE TO BUSINESS FOR ENGINEERS AND SURVEYORS

A POCKET GUIDE TO BUSINESS FOR ENGINEERS AND SURVEYORS

H. Edmund Bergeron, P.E., MBA

JOHN WILEY & SONS, INC.

This book is printed on acid-free paper. ♾

Published by John Wiley & Sons, Inc., Hoboken, New Jersey
Published simultaneously in Canada.

Library of Congress Cataloging-in-Publication Data:

Bergeron, H. Edmund.
A pocket guide to business for engineers and surveyors / H. Edmund Bergeron.
p. cm.
Includes index.
ISBN 978-0-471-75849-5 (cloth)
1. Engineering management. 2. Business ethics. I. Title.
TA190.B47 2009
620.0068–dc22

2009009725

Printed in the United States of America

SKY10032416_010722

CONTENTS

PREFACE

I'm sure that many readers have suddenly found themselves thrust into a position of managing a project team, interfacing with clients, and writing proposals to obtain the next project. Engineers and surveyors who find themselves in this situation generally are excellent technicians who have been called upon by their boss to take the next step up the ladder toward becoming a project manager, group leader, associate, principal, or even a partner in the firm. Perhaps you even sought an increased level of responsibility because you knew it was necessary to advance your career.

There are several potential problems with this situation. First, and probably most important, is that you've had very little formal training in the many facets of business. That's probably why you are reading this book.

Engineers and surveyors are smart and pick up new things very quickly, and many bosses think learning about business topics such as communications, human resources, marketing, or accounting can be learned on the job similarly to learning to use a new piece of surveying equipment, new technical software, or a new code or standard.

Second, while you may be a very skilled technical person, you may not have the skills or the natural inclination needed to become a good manager. It has been said that "young engineering and surveying professionals are promoted for their excellent technical skills and subsequently fired for their lack of management ability." I've seen it in my own firm, and hopefully this book will help you avoid this pitfall.

Finally, engineers and surveyors have such a great love for the technical aspects of their work and the desire to produce the "ultimate" technical solution that they practice their profession first, and business second. Believe me, this is not how the rest of the world does it. Have you ever been admitted to the hospital before presenting your insurance card or signing a bunch of waivers? Have you ever had a lawyer or accountant provide services to you before signing an agreement and providing them with a retainer fee? Does your firm often waive interest on overdue accounts because you feel guilty asking for it? Well, many engineers and surveyor do just that. We are so enthralled by the project that we don't furnish our client with a sufficient fee estimate and agreement *before* beginning the work. Sometimes we aren't even clear about the scope of the project or even who our client is.

If some (or a lot) of this sounds familiar to you, *A Pocket Guide to Business for Engineers and Surveyors* is for you. This book isn't an MBA under a single cover, but it will provide you with a reference for your most frequent business problems. Throughout the book, we'll cover the management topics that are so essential to a successful career for you and the long-term sustainability of your firm. This book is primarily intended for young professionals with 4 to 8 years of experience who have just received their license and are beginning to think about the management aspects of their career. With the technical hurdle of licensing cleared, you may be thinking about your career in the long term, which will inevitably include management responsibilities of one degree or another. Seasoned professionals and new business owners also will find this book helpful. If you've had some formal management training, it will serve as a quick reference when unique problems arise, and if you haven't, it will assist you in developing solutions to your most common business problems.

Remember, there is a primary difference in mindset between being an excellent technical engineer or surveyor and a successful businessperson. Excellent engineers and surveyors generally evaluate several alternates to find the best solution, then we analyze plenty of data to verify the measurements and support the conclusion. Generally, we take a "belt and suspenders" approach to every technical problem. Given the high degree of liability and responsibility for public safety inherent in our profession, this is understandable and is absolutely necessary. I'm not suggesting that, because you've begun to take on some business responsibility, you should change your technical approach one bit, but business problems seldom provide all of the data or allow time to evaluate many alternates. Business decisions are often required *now* and require you to make them with less than complete data or evaluations of all the alternates. This is reasonable, since a poorly made business decision doesn't have the grave circumstances that a miscalculation would cause in a technical problem. Generally, a poor business decision isn't a public health and safety issue, but it can negatively impact your business and the ability of the firm to meet future obligations and challenges.

I hope you find that this book meets your expectations as a daily practical reference for your most common management problems. Good luck to you in your future as a manager in our exciting profession!

1

PROFESSIONALISM & BUSINESS

"We've got a great profession to practice but a lousy business to be in."

I used to teach a graduate course entitled "Issues in Engineering Management and Practice." Now I teach it as a seminar series in Engineering Management, which is attended primarily by practitioners. In both cases the preceding quote is the first thing I discuss with students. Why? Because most young engineers and surveyors are optimistic idealists who derive self-satisfaction from solving complex technical problems. Yet they are naïve when it comes to the business aspects of a successful firm, earning a client's respect, and effectively working with others. Overall, surveying and engineering schools have done a great job of turning out excellent technicians who have had little training in seeing the "big picture," solving people problems. A high level of idealism and optimism is certainly healthy. It is required to create innovative solutions for our clients and successful projects. Understanding who we are as professionals, how the business of professional practice interacts with idealism and our client's problems, and where we fit in within our firm (or should we risk going out on our own?) are all very important parts of being a professional surveyor or engineer. My personal, and possibly over simplified, definition of a professional versus a technician is: a technician solves technical problems and a professional solves people's problems. This chapter will explore various characteristics of being a professional, the difference between a technician and a professional and how a professional interacts with the business of professional practice.

WHAT IS A PROFESSIONAL SURVEYOR OR ENGINEER?

How do professionals compare with a nonprofessional? It is generally understood that professionals, whether doctors, dentists, attorneys, accountants, architects, engineers, or surveyors, has advanced education and experience and is licensed to practice by the state where they reside and other states where they may also practice. For engineers and surveyors, an advanced education is not required by most state licensing laws, although many have obtained additional education beyond their primary degree. The National Council of Examiners for Engineers and Surveyors (NCEES) has developed model laws for both professions, which are used by most states as the basis of state statutes that regulate practice and assure the public that minimum standards have been met in order to ensure the public health and safety. These laws state that professional practice is based on a "three-legged stool." The first leg is an ABET-EAC (Accreditation Board for Engineering and Technology—Engineering Accreditation Commission) accredited bachelor degree for engineers and for surveyors an ABET-TAC (Technical Accreditation Commission) associate's degree or an ABET-EAC bachelor degree. The second "leg" is examinations. Both engineers and surveyors are required to take two levels of examinations. Engineers generally take the Fundamentals of Engineering Exam, or FE, exam near the end of their junior year in college, fall of senior year, or shortly after graduation. Passing this exam gives the graduate the right to use the title "Engineer-in-Training," designated by "EIT" after their name or, in some states, "Intern Engineer" with the designation "IE." Surveyors take the Surveyor in Training exam upon graduation or shortly thereafter. Surveyors who pass the exam may use the designation "SIT" after their name. Neither the engineer nor surveyor has the legal right to individual practice after passing the initial exam. They still must continue to gain experience by working under a licensed professional. Members of both professions take a second exam upon obtaining additional experience "satisfactory to the board." Experience is the third "leg" of the stool. For an engineer, this means a minimum of four years of experience "of progressively increasing levels of complexity and responsibility," while being overseen by a licensed engineer supervisor. In some cases, a state board of licensure may require one year or more of additional experience if they deem that the engineer's initial year of experience to be not at a professional level or increased level of complexity. This may happen to a young engineer employed in government or construction but rarely happens to those employed by consulting firms. Surveyors are required to have either two or four years of experience, depending upon whether they posses an associate's or bachelor degree. Once either professional has the required experience, they may sit for the professional part of their exam. The engineer's exam is known as the Principals and

Practice exam and the surveyor's exam is known as the Professional Practice and Legal Exam. After successfully passing the exam, engineers become professional engineers and may use the designation "P.E." Surveyors upon passing their exam become licensed land surveyors with the designation of "LS" or "LLS," depending on the state where they practice.

The previous paragraphs explain the technical and legal requirements for becoming a professional, but how does the public determine who is a professional and who isn't? First, most professionals refer to those who purchase their services as "clients" not "customers." A subtle difference, but client implies personal service and attention given to individuals. Customers are all generally understood to be all the same and receive only minimal personal service. Clients may be individuals, companies, or government agencies. The term client implies a higher level of respect, understanding, and concern for the problem that needs to be solved. It also implies a high level of service and attention to the client's needs, since the professional possesses a high level of technical skill and knowledge that the client generally doesn't have or, in many cases, doesn't even understand. The term client also implies a level of trust that customers generally don't have with the provider of goods and services.

The attributes of a profession have evolved over the years. In the July 1957 issue of *Social Work*, Ernest Greenwood, Professor Emeritus, UC Berkeley, published a paper describing the five attributes of a profession as:

Systematic theory
Authority
Community sanction
Ethical codes
A culture

While Greenwood's paper is over 50 years old, it is my opinion that his five attributes remain the foundation for describing a profession. More recent writing has only built upon and amplified these basics.

- *Systematic theory*. Each profession has a systematic theory as its basis, upon which the profession is defined. The engineering and surveying professions define themselves as being based upon theories of mathematics and science. Educational accrediting agencies such as ABET prescribe specific criteria as the educational foundation for practice of the profession. More recently, the American Society of Civil Engineers, in 2005, published the first draft of the Body of Knowledge, which, in addition to ABET criteria, prescribes 15 criteria as the basis for professional practice in civil engineering.

- *Authority*. Greenwood states that "In a professional relationship, the professional dictates what is good or evil for the client, who has no choice but to accede to professional judgment." While much has changed since 1957, including an increase in public skepticism of professionals, nonetheless professionals do have advanced and specialized knowledge upon which their clients depend in order to resolve their problems.
- *Community sanction*. This is the right, some say the obligation, of the profession to govern itself. This is evidenced in all states through professional licensing laws and licensing boards. The premises of the law being to "protect the public health and safety." In recent times, the public skepticism of professions effectively regulating themselves has been shown in licensing boards being required to have "public members" on the boards in order to represent the public interest.
- *Ethical codes*. Surveyors and engineers have codes of ethics that were adopted by their respective professional societies. These govern professional-client relationships and professional-professional relationships. Provisions in codes of ethics that restricted professional business practices and fees were struck down in the 1978 Supreme Court decision *National Society of Professional Engineers v. United States* as violations of the Sherman Antitrust Act. Many state licensing laws have incorporated codes of ethics into their licensing law in an effort to improve the ethical behavior of engineers and surveyors. Overall, professionals and the public believe that codes of ethics are not uniformly enforced and have little effect on unprofessional behavior.
- *A culture*. Greenwood also states that a "profession culture" exists. It consists of social values, norms for professional behavior and symbols of practice. He says that social values are shown in the effort of the professions to regulate themselves for the public good. Norms of professional behavior now are largely defined by court decisions as to what does or does not constitute a "standard of professional practice." Professionalism is also defined by symbols that constitute history, professional dress, insignias or emblems, and profession jargon or buzzwords. Engineering symbols, generally known to the public are a slide rule, drafting scales, triangles, and more recently computers. The most common surveying symbols are the compass, transit and steel tape, and drafting equipment. Each engineering and surveying society also has its own logo or insignia.

Dealing with Clients in Trust and Business

Who are the clients that professional surveyors and engineers provide services to? Both professions have many clients. These can be private individuals who engage the professional on a one-time basis to provide a boundary survey of their home lot or it may be an engineer to design a septic system or investigate a leaky roof or cracked foundation. The owner is truly in the position of needing to trust the professional, since it is unlikely that he/she will have detailed knowledge of the service being provided and sometimes even of the problem to be solved. Clients also can be developers who engage engineers and surveyors regularly and often engage the firm in a position of trust for project after project. These clients are knowledgeable and sophisticated regarding the professions and the services needed. Clients also can be large corporations or government agencies, where the professional is engaged after the evaluation of qualifications or submission of a competitive fee. In this instance, the board or committee that hires the professional may not be the client contact or the direct recipient of the professional's services. The ultimate recipient of the services may be a sophisticated professional such as a staff engineer or surveyor or an experienced maintenance director who understands the scope of work and the complexity of the problem and empathizes with the engineer or surveyor. The client also may be a manager or director who has little interest in the details and only wants to know that the problem is solved or the project is complete. There may be little or no opportunity here for establishing a long-term relationship, and the type of agreement may make it difficult for the engineer or surveyor to establish a high level of trust with the client. I've also found, through my years of practice, that professional engineers and surveyors deal with their clients very differently from the way companies deal with customers. Many engineers and surveyors, especially sole practioners or those in small firms, are so interested in the technical challenge of the problem to be solved or the social benefits of the project that they begin and sometimes even complete an assignment before they even discuss the business arrangements or fee with their client. In fact, the New Hampshire Board of Licensure for Land Surveyors states that one of the major complaints against surveyors is not technical competency but a lack of clear business communication by the surveyor in terms of fee for services for a particular project. Until I'd been in private practice for 10 years I also thought that doing the project and discussing payment sometime later, often after the work was done, was they way the entire business world did business. When I decided to return to school to obtain an MBA, I learned that the rest of the business world did things quite differently. I was surrounded by business professionals from many other industries and practices and quickly learned that businesses outside of the practice of engineering and surveying work out the details of the business arrangement in terms of what will be done, when it will be done, and how much it will cost *before* they begin the work. If the fee or product cost is too high, the customer goes elsewhere. John Bachner, a noted author on engineering professional practice says "It is the design professional's lack of business acumen, not greed that creates so many of their problems." Bachner also states, "It is not unprofessional to lack business skills. It is however unprofessional to recognize a shortcoming but then do nothing to correct it." How true that is!

Professional Practice Is Not 8:00 to 5:00

How else are professionals distinguished from nonprofessionals? There is little discussion regarding the work ethic of professional surveyors and engineers in professional literature and even less discussion during our professional education. When I graduated from university many years ago I went to work for a state department of transportation. The standard work hours were 8:00 AM to 5:00 PM with an hour for lunch. I never quite caught on to the 8 to 5 routine, since often I'd be working on a challenging design problem, which I was compelled to finish before I went home, or I found it more efficient to check the work of technicians working on my project in the quiet evening or early morning when no one else was around. I also felt that deadlines were important and my boss expected them to be met. If it required a little more time to meet a submission deadline, I'd put in the time necessary to make it. I was a salaried employee, not hourly so my employer received the direct benefit of my extra effort. Some of my fellow engineers didn't see it that way. They'd begin winding down, clearing their desks, and rolling up their plans at ten minutes of five, and you didn't want to be coming in the drafting room door as the clock struck five and the masses were exiting. Now most of these fellows (in those days there weren't any women engineers) were good engineers, but were they really professionals? Does the surgeon leave his patient on the operating room table when it's time for lunch or does an attorney stop his cross-examination when the clock strikes five o'clock? Needless to say, I didn't find the DOT atmosphere conducive to my definition of professionalism and after a few years I moved into private practice.

It is my contention that you can't teach professionalism or a professional work ethic in surveying or engineering schools. Many programs now do a much better job trying to simulate real-world problems, issues, and clients now than when I was in school. Most programs require students to do at least one group project per semester. Usually these projects are closely tied to real-world projects, and the student groups are required to define different roles and responsibilities for each member of the group. Generally, they also are required to make presentations to faculty and sometimes practitioners or interested members of the public. This is fine, but I recently had to remind a summer intern that wearing his baseball cap and blue jeans to a client presentation wasn't up to the level of professionalism that our clients expect. Young engineers and surveyors in our firm quickly learn that clients pay their salary, and close attention to personal appearance, client projects, schedules, and first class client service are always required. Project managers in our firm, who are licensed professionals, often take young engineers to evening meetings where their projects are presented to regulatory boards and the public. At these meetings, the personal appearance of a professional, appropriate to the local custom, is required if you want to earn the respect of the approval board or agency and effectively represent your client. Surveyors in our firm also understand that if they are working on a survey that is almost done at 4:00 PM, they need to stay in the field at little longer and finish the assignment rather than spend an additional trip the next day for an hour and a half of field work. *Professional practice in the engineering and surveying profession is not an 8 to 5 job!*

THE DUAL CAREER PATH

Does a dual career path really exist? One for those who are only interested in the technical side of their profession and one for those who find the challenges of managing people and projects to be just as interesting as the technical side of the project? Can you be a real professional and only interested in the technical aspects of a project? I often hear young surveyors and engineers say that they chose their career because of their preference for mathematics, science, wanting to work outdoors, and the self-satisfaction received from solving technical problems. In other words, "I'm not interested in sitting at a desk eight hours a day, or I really love CAD drawings and solutions I can produce at my new work station but don't bother me with schedules, budgets, and all that client stuff." Louis Berger, former chairman of Louis Berger International and recipient of the ASCE, Parcel-Sverdrup Award in engineering management in 1995, stated in his acceptance speech that engineering managers currently make up 22 percent of the 2500 people employed by Louis Berger in 80 countries throughout the world. These include 2 percent of their staff who run independent offices, 4 percent who are capable of managing large projects, 6 percent who manage small projects, and 8 percent who run small design teams of 5–10 people. He believes that firms of the future, both small and large, will require an increasing number of engineers who have excellent management skills. He uses the example that many clients now insist that presentations at interviews where the firm is selected be given by the team who actually will work on the project. This requires that project managers and senior engineers have excellent communications skills both written and verbal. Berger stated that "regretfully, it also eliminates some civil engineers who can not present themselves well verbally." The answer to the dual career question is a definite yes, but it also becoming clear that those engineers and surveyors who bring work into the firm and successfully manage people and projects will progress higher up the corporate ladder and receive larger salaries over their lifetime. Since surveying and engineering curricula put very little emphasis on the soft skills required of management, surveyors and engineers must learn these on the job, through self-teaching or through additional formal education in management such as an MBA or master of engineering management degree. Those who elect to stay in the technical area of the profession must also continue to increase their technical expertise if they are going to move upward in their career. This also will require additional education and training, often via a formal master of engineering or surveying program. Typical job titles on both career paths in our firm are as follows:

Technical Career Path

- *Entry level.* Junior engineer (EIT) or survey instrument person (SIT)
- *First step/promotion.* Project engineer (EIT) or survey part chief (SIT)
- *Second step/promotion.* Senior project engineer (PE) or project surveyor (LS)
- *Third step/promotion.* Chief engineer/technical leader (PE) or survey department leader (LS)

Management Career Path

- *Entry level.* Junior engineer (EIT) or survey instrument person (SIT)
- *First step/promotion.* Project engineer (EIT) or survey party chief (SIT)
- *Second step/promotion.* Senior engineer/project manager, small projects (PE) or senior surveyor/project manager, small projects (LS)
- *Third step/promotion.* Senior engineer/project manager, large projects (PE) or senior surveyor, large projects (LS)
- *Fourth step/promotion.* Senior engineer/department head (PE) or senior surveyor/department head (LS)
- *Fifth step/promotion.* Principal engineer (PE) or principal surveyor (LS)

We also hire young engineers and surveyors to work as summer interns (usually the summer between junior and senior year). This gives us an opportunity to evaluate young professionals and determine if we'd like to offer them a job upon graduation. It also allows the interns to "try out" the profession to see if this is really what they want to do for the rest of their career. We've had some interns who have determined that consulting engineer and the engineering profession is not for them. It is much better for them to find out before their senior year and graduation than after they've graduated.

As you can see, the technical career path maybe somewhat limited in a small consulting firm, and the management career path can lead to a higher-level position. It is not that consulting firms reward those on the management path more than those on the technical path, both are needed for a firm to be successful. Those who provide a high level of client service and bring in new work are essential to survival of the firm. Either career path can lead to ownership within a firm that recognizes the important contribution of both paths. Ownership will be discussed in a later chapter.

BEING AN ENTREPRENEUR VERSUS CLIMBING THE CORPORATE LADDER

Trying It on Your Own

Some surveyors and engineers feel that they are working hard for the benefit of their clients but their efforts are not being recognized by their boss or the firm's principals. Some feel that if they are putting in 50–60 hours per week and they should reap some of the benefits and profit themselves. Some feel that coworkers are doing less than their "fair share" and are causing a drain of the firm's profits by constantly finishing projects over budget and behind schedule. Others feel that they have technical and leadership expertise, and others recognize them as the "go to" guy when the going gets tough. No matter how you put it, some of these individuals feel as though they are getting a raw deal and may consider starting their own firm. What are the key characteristics necessary to head out into the uncharted waters on your own? The first is that you need to be a risktaker, an optimist, and have a mission. When I started my firm 35 years ago my mission was to control my own future and to create a small consulting firm that would provide basic civil engineering services in a rural community where none had existed in the past. I believed I had the technical expertise, being both a licensed surveyor and engineer, as well as the desire to make a difference in the community where I lived. I didn't have a business plan, any clients, office space, or equipment and had only a crude market survey that showed a reasonable chance of success if I worked hard. Was I a risktaker or just idealistic and naïve? The answer didn't come for many years and, to be honest, I don't think it made any difference. What really counted though was the desire and ability to work hard 24/7 three hundred sixty-five days a year for the first few years, along with incredible optimism, shared by my family, our staff, and clients that the firm was having an impact in the community. Good people and communication skills, both written and verbal, also helped. I also took the opportunity to learn as much as I could from other successful engineers whenever I had a chance.

Another important consideration for starting your own firm is how do you finance a start-up firm? Lack of start-up capital is an almost universal problem with all small businesses. It causes most firms to grow at a rate much slower than market forces may allow. It was much easier when I started than it is today. We had no mortgage payment, since we rented our house. My first office was also rented space, so capital outlay for real estate was the security deposit. In the "old days," the only significant investments were a drafting table, survey instrument, steel tape, book of trig tables, and a calculator. The family car doubled as a survey vehicle, and there was no expensive investment in copiers, CAD workstations, networks, or sophisticated software. My first professional liability insurance policy cost $200! I started working mostly by myself without health insurance for our young family and contracted help when I needed a survey crew. Things have changed, and if you are planning to try going out on your own, develop a simple business plan, including where the work will come from and your expected start-up cash flow and have significant savings (preferably enough to last a year without any income) or a rich uncle. I have heard of the owner of a recent start-up firm applying for as many credit cards as he could get and then maxing them out in order to fund his start-up firm. Obviously, this was very risky, but it probably was the only way he could manage, since bankers and investors seldom lend money to start-up firms based on the owner's mission and commitment to hard work. Also, it doesn't hurt to have a spouse who has a fairly secure job with a good salary and health benefits.

Climbing the Corporate Ladder

If risktaking and incredibly hard work with no pay and no benefits aren't for you, consider what it takes to climb the corporate ladder within your present firm. What are your personal goals? Can you not stand to be without income for significant periods of time? Do you want to work 40–50 hours per week, have most weekends to spend with your family, and attend only an occasional evening meeting? Do you want to be a principal, leader, and owner of your current firm? Who is your competition and do you have what it takes? If the answer to many of these questions is yes, and you want to advance up the corporate ladder in your current firm, you many be in for almost as much hard work as you would if you started your own firm but with a salary and benefits and at considerably less risk. The most important way of advancing in your firm is to distinguish yourself from the rest of the pack. Become known as the "go to" guy or gal that the boss can depend on in any situation. A friend once said, "If you're not well educated, it pays to be versatile." Some other key characteristics may include:

- Develop a leadership style that earns the respect of your clients and coworkers.
- Find a high-level mentor in your firm and emulate the behavior that brought him/her success.
- Communicate frequently with your boss and coworkers regarding client, firm, personal, and project goals.
- Take on unpopular or difficult assignments joyfully and accomplish them successfully.
- Learn how to delegate and accomplish much more than you can alone by utilizing the skills of others on your team.
- Establish a work ethic that gets assignments done with minimum follow-up required by your boss.
- Insist that your team only provide a high-quality work product.
- Provide first-rate client service on time and within budget.
- Bring in new work from existing clients, and play a key role in marketing new clients.
- Obtain an advanced degree in surveying, engineering, or business.
- Be a leader in your community outside of the firm. Serving on non-profit boards, coaching the little league teams, being a scout troop leader, and/or volunteering in the church or local soup kitchen will enhance your personal reputation and could even bring work into your firm.
- Give back to the profession through participation in professional societies and organizations. Start as a committee member, progress to chairperson, and go on to become an officer.

Does all of this sound like hard work? It is, but the reward can be prestige, financial success, and advancement to a high level in your firm with minimal risk compared to heading out on your own.

TYPES OF BUSINESS ORGANIZATIONS AND THEIR STRUCTURE

Legal Forms of Business

What legal types of business organizations are most common for surveying and engineering firms? How simple or complicated are they to set up? What are the advantages and disadvantages of each form of organization? A number of types of engineering and surveying firms exist, the simplest being a sole proprietorship and the most complicated being the corporate form. The different forms are:

- *Sole proprietorship*. There is no formal legal process to set up a sole proprietorship. It is one of the most common forms used when individuals start a new business. Taxes for a proprietorship are paid as part of ones individual tax return using IRS Schedule "C," which shows profit or loss from a business. Obviously, the proprietor is the owner and boss and keeps all of the profits as well as taking any losses. This can create difficulty with employees, promotions, and future ownership transition.
- *Partnerships*. Partnerships are made up of two to hundreds of people. Many older firms, particularly law firms, started as partnerships and some continue that form of business today. Partnerships can be risky, since in an equal partnership all partners share, risk liability, profit, and loss equally. This can mean both inside and outside of the business. A partner's individual financial problems could become a problem for the rest of the partners. For this reason, partnerships are no longer a popular form of business, and partnerships that are set up generally are limited partnerships, which limit the responsibility of each partner for the liability of other partners. Partnerships also pay taxes through an individual tax return. They do, however, have to file a tax form that notifies the IRS of the partnership income, assets, and liabilities. Each partner, in turn, receives a K-1 form, which reports their share of the partnership income or loss, and it is attached to their tax return. Partnerships also can create difficulties for employees. An employee who doesn't progress to the position of "partner" in a prescribed number of years often is considered to be a lower-level achiever and may leave the firm. Ownership transition is also difficult in a partnership, since there is no stock to transfer.
- *Corporations*. Corporations are created as legal entities to limit liability and to create stock, which, theoretically, is a liquid asset that makes transfer of ownership easier. Corporations also can borrow money and own assets in the name of the corporation. There are many types of corporations and some types are controlled by the laws of the state where they are set up.
  - "C" corporations are the traditional form of corporation that most people are familiar with. They can be privately owned or the public can own them by trading stock on the market. The corporation files a tax return and pays corporate income tax on its profits or takes related

losses. Many surveying and engineering firms are organized as "C" corporations. Most firms make an effort to pay out most of their profits at year end by giving bonuses to principals and employees in order to avoid double taxation and minimize overall tax liability of the corporation. Ownership in a "C" corporation is available to anyone who purchases stock, so ownership transition is easy with a "C" corporation.

- "S" corporations are similar to partnerships in that they pay no corporate income tax and the profits flow to the stockholders where they are taxed at the individual level. They have the advantage of limiting liability to the corporation itself. "S" corps have a limit on the number of stockholders which may present an issue at the time of ownership transition. Many small surveying and engineering firms are organized as "S" corps.
- LLCs, or limited liability corporations, can be similar to "C" or "S" corporations in structure. The principal reason for setting them up is to specifically limit liability to the assets of the corporation itself. They may not be very effective for surveying and engineering firms, since laws governing professional practice in many states allow "piercing the corporate veil" in order for liability to flow directly to the licensed professional who is ultimately responsible for the work.
- "PAs" or "PCs" are professional associations or professional corporations organized under the specific business provisions of the professional licensing laws of many states. The licensing laws usually require the principals and/or stockholders of the corporation to be licensed professionals. For tax purposes, they can be "C" or "S" corporations. This may present some problems when a transfer of ownership to stockholders who are not licensed professionals is proposed.

Before deciding which form of business is appropriate for your firm, consult an attorney and tax consultant.

Business Structure

The operational structure of a business is independent of the legal form of business and generally relates to the type of client or service a firm offers. The structure often changes as a business grows, in order to take advantage of unique capabilities and interests of individuals. It also often changes when ownership changes. The traditional form of operational structure for a medium-sized engineering firm of 20 to 50 employees includes a president and various vice presidents, as shown in Figure 1.1.

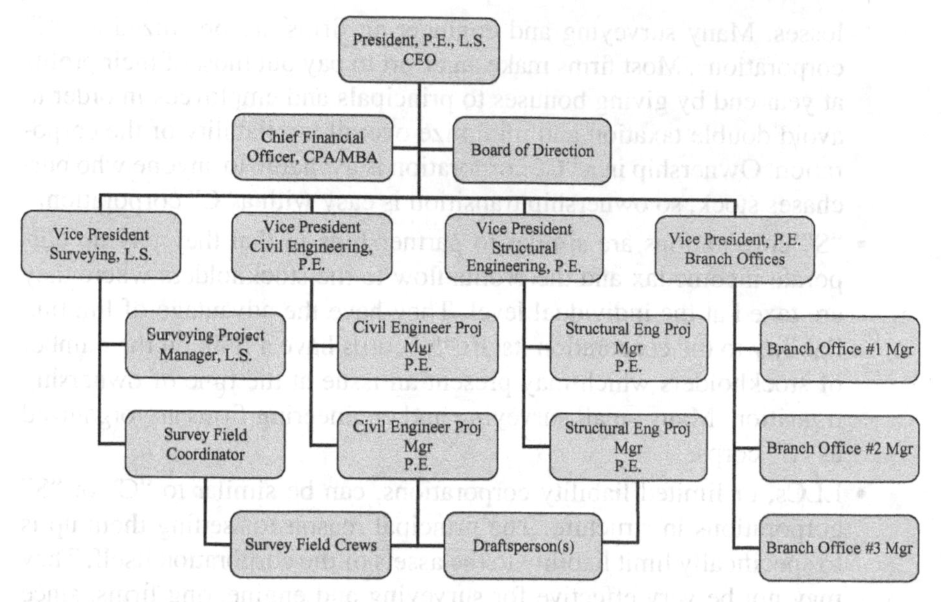

Figure 1.1 Organizational chart for a traditional medium-sized firm.

A simpler organization for smaller firms of 5 to 15 employees also follows. Note that in a small firm the same person may serve in more than one position, as shown on the organization chart in Figure 1.2.

Many firms today are organized as a type of organization that functions according to the needs of their clients. These organizations may offer services similar to those of a traditional form of business, but they are primarily

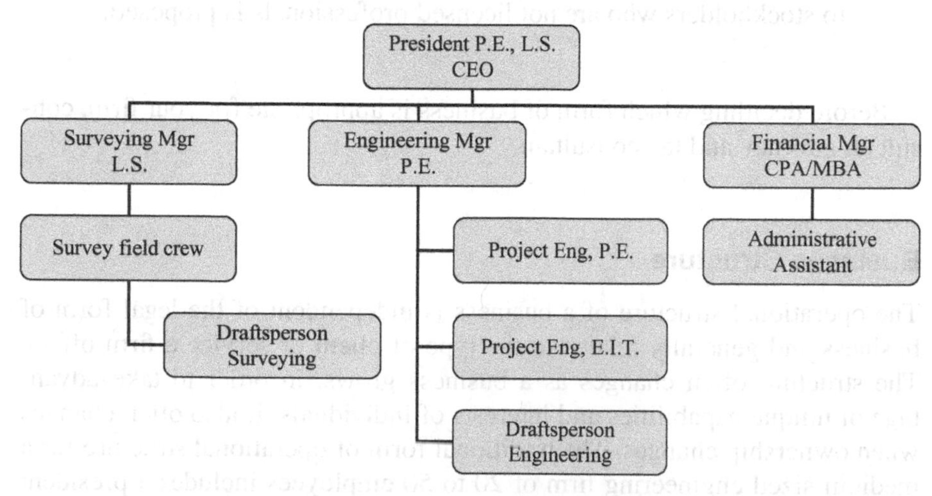

Figure 1.2 Organization chart for a traditional small firm.

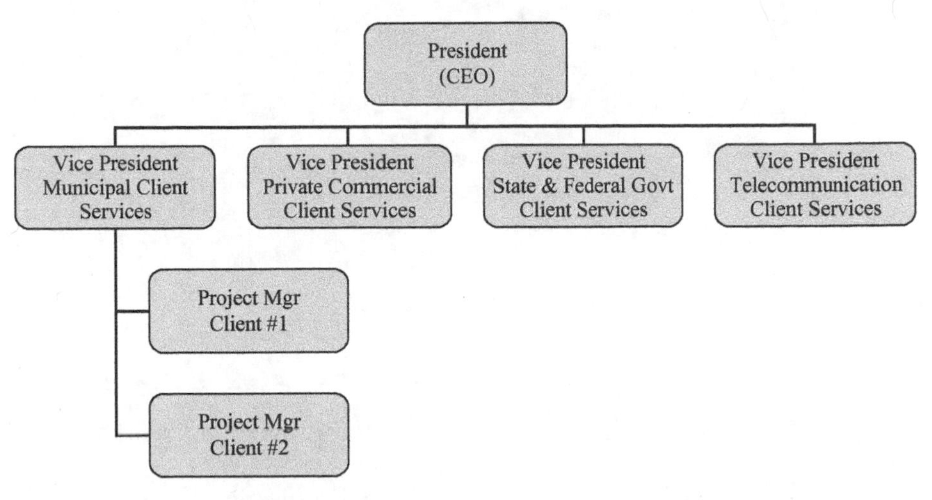

Figure 1.3 Client service–based organization.

Table 1.1 Structure of Organizational Matrix

Projects Resources	Project #1	Project #2	Project #3
Survey	•	•	
Geotechnical	•		
Wetlands	•	•	•
Environmental	•		•
Civil staff	•	•	
Drafting staff	•	•	
Administration	•	•	•
Construction insp.		•	
Owner	•	•	•

organized to serve specific types of markets or niches, as shown in Figure 1.3. Table 1.1 illustrates the breakdown of responsibility for three projects in a matrix organization. Engineering and surveying organizations are one of the last types of service organizations to change to client service organizations, which focus on client markets or niches.

Communication technology today allows firms to organize using a virtual organizational matrix for a single client or a specific project. In this arrangement, individuals contribute their expertise for a single project or client and then may move on to another organization for the next project or client.

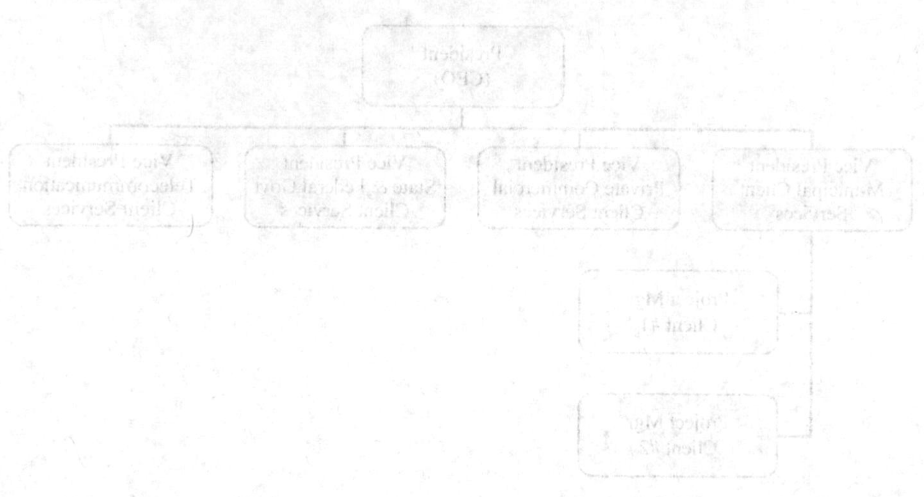

2

MANAGING YOURSELF

> "If you always do what you've always done, you'll always get what you've always got."
>
> —Sally McGhee

> "The significant problems we face can't be solved at the same level of thinking we were at when we created them."
>
> —Albert Einstein

Who am I and how do I affect my personal actions and the actions of those around me, and is it necessary for me to change? Do you feel as though you should loose 10 pounds but can't stick to a diet and shun exercise? Were you the type of student who always knew you could have achieved better grades if you only studied a little more? Would you have been a better athlete or musician if you had practiced with more dedication instead of relying on your natural ability? Are you going to learn a new skill, hobby, or foreign language, when you have time? Would you like to be a leader in your community or company but need to work on your people and business skills?

I'm a procrastinator. We all are to a certain extent, especially with those things which we find particularly unpleasant to do. Generally, we avoid doing things we don't like to do or those that take too much time or are harder than the average task or that we have less than the desired aptitude for. We often rationalize procrastination by saying, "My life is too busy," "I don't have enough time now," "There are not enough hours in the day" or "I'll get to it as soon as things slow down a bit." However, have you noticed how you seem to be able find a way to fit into your busy schedule: socializing with friends, a game of touch football, an hour for your favorite TV show, or some other task or pastime that you really enjoy? What about basic needs like eating, sleeping, and personal hygiene? We generally find time to do the things we enjoy or need to do to survive.

I've been known to paraphrase Sally McGhee's popular quote in a little less formal terminology, "If you're gonna do what you've always done, you're gonna get what you've always got." Several years ago my brother put it to his son, then in his late teens, this way, "if you're a dubba in your twenties, you're gonna be a dubba for the rest of your life." You get the idea! If you want to change your personal situation, you have to change yourself first. Realize that *you* are the problem! It sounds easy doesn't it but most of us have a pretty good life as it is so why change? Inertia is a very hard thing to overcome.

In this chapter, we'll first try to understand ourselves, our beliefs, and what can motivate us to change. Then, we'll determine if change is necessary and how to do it. We'll explore deep-seated beliefs, or paradigms, since changes in these are necessary to institute real and lasting change. This chapter will also help you better organize, through time management and controlling interruptions, to be more productive and effective. We'll also explore Maslow's Hierarchy of Needs to provide a basic understanding of physiological and self-needs that must be understood and satisfied if we are going to invoke personal change.

A PERSONAL QUIZ

Before we begin an in-depth study of ourselves and our beliefs, let's look at some reasons why you may want to change the way you do things.

- Do you get out of bed unexcited about going to work each day?
- Does your day consist of fighting "brush fires" and managing the next crisis?
- Do you feel as though your life is out of control?
- Are you continually interrupted by others?
- Do you have too many unanswered emails?
- Do you have too much information to deal with?
- Is your "to-do" list growing and is nothing getting done?
- Do you spend too much time looking for things?
- Do you feel as though planning for tomorrow makes no sense because it's only going to change?

- Do you feel as though there isn't enough time in a day?
- Are you disorganized by nature?
- Do you feel that forced organization and planning will cramp your style?
- Do you often wonder what you've done at the end of the day?
- Do you go home feeling as though you haven't accomplished what you needed to do today?
- Are you stressed because you work too hard and don't have enough personal or family time?

If your answer to some, or many, of these questions is yes, you need to be proactive in managing yourself and the way you do things. The result will be a happier, more productive self with a high level of self-esteem and self-satisfaction.

PARADIGMS

Deep-Seated Beliefs

What is a paradigm and how can it affect change in your behavior? Paradigms are deep-seated beliefs, theories, and frames of reference that we believe to be true. They generally form the basis of much of what we believe and how we behave. They are the way we see things although they are not natural laws such as Newton's laws of physics. For example; the sun always rises in the east and sets in the west. This is a natural law, or law of physics. Force equals mass times acceleration, or $F = ma$, is Newton's second law. The following are examples of paradigms. It's a home run when a baseball player hits a baseball over the "Green Monster" at Fenway Park. Columbus sailed to the New World and arrived thinking he'd found the West Indies. In the United States, we drive on the right side of the road, and in the UK they drive on the left. As you can see, paradigms are not just common beliefs; they are universally believed to be true, but unlike natural laws they can be changed. Our paradigms are based on deep-seated principles that we have believed to be true over a long period of time. Although it has been true since the days of Babe Ruth, a ball hit over the famous Fenway "Green Monster" could be declared a double or triple by changing the rules. In the UK, people could drive on the right side of the road, but it would take major retraining of British society and considerable additional work to change road signs and redesign traffic lights and intersections.

How do our paradigms develop and how do they influence the way we think and do things? Stephen Covey in his book, *The 7 Habits of Highly Effective People* (Free Press, 2004), uses the example of a pencil sketch of a woman to demonstrate our bias toward certain paradigms. The test was first used at Harvard Business School to demonstrate that two people could look at the same picture and see different things, disagree about what they saw, and both still be right. Half of the class was given a sketch of a beautiful young girl which they were allowed to look at for only 10 seconds, and the other half of the class was given a sketch of an ugly decrepit old woman and also allowed to view it for only 10 seconds. A composite sketch was then presented to the whole class. When asked what they saw, students in one half of the class saw a decrepit old lady, while the other half saw a beautiful young girl. What appeared to one half of the class to be the old lady's nose, mouth and hooked chin was viewed by others as the cheek, necklace, and slender neck of the young girl. Lively discussion ensued where students described what they saw to each other and tried to convince those who saw the opposite that what they saw was the truth. The exercise demonstrated that what we believe, our paradigms, are based on our experiences and deep-seated beliefs, that they may be completely different from what others see, and that both of us may be correct. This may seem impossible, but the way we see the world is based on our paradigms, our core beliefs. Our paradigms are our deep-seated beliefs, but they may not be those of others, our friends, fellow workers, competitors, or enemies, and both of us can be correct. The important thing to understand about paradigms is that if we are going to change significantly, it is going to take changing our core beliefs, or as Covey states, a "paradigm shift."

In order to demonstrate paradigms, try the beautiful young girl/ugly old woman test yourself. Look at each of the three sketches that follow. The first sketch is of a young lady looking away from us. You may want to look at this one while you ask a coworker to look at the sketch of the ugly old lady. Can you see her bonnet, the profile of her forehead, her petite nose and a bead necklace? Look at this sketch for no more than 10–15 seconds. Next, have your coworker look at the sketch of the old woman for the same amount of time. What do they see? Do they recognize her big nose, deep set eyes, mouth, and sagging chin? Finally, each of you look at the composite sketch. Which woman did you see first? Which did your co-worker see? Did you see the beautiful young girl with a petite nose, and a pearl earring, who now appears to be wearing a woman's hat with a feather in it and a black shawl, or did you see the ugly old woman with a large nose, tense mouth, protruding chin, and an overall sad look on her face? The first woman you saw was very likely affected by your paradigm, the first sketch you looked at. If you saw the beautiful young woman first, you probably had to look really hard to see the ugly old woman. You can try this exercise with several friends. Show half of group the sketch of the beautiful young woman first. Show the other half the ugly old lady first. Be sure to limit their viewing time to 10 to 15 seconds. Next, show each group the composite sketch and ask which woman they see

Figure 2.1 Sketch of young lady.

first. Have each group try to convince the other that what they see, the woman in the composite sketch affected by their paradigm is correct. This exercise graphically demonstrates our paradigms, how they are affected by our core beliefs and how much work it takes to change them. Remember this exercise. It probably will not be this easy to change your paradigms in real life.

Figure 2.2 Sketch of old lady.

Figure 2.3 Composite sketch.

PRINCIPLES

Personality Ethic versus Character Ethic

In addition to understanding our paradigms, Covey feels that our core beliefs are affected by our ethics. At the time he was writing *The 7 Habits*, he also was researching 200 years of writings of success literature from 1776 to 1996. What he found was that from 1776 to the period before World War II most scholarly discussions on success and self-improvement emphasized traits such as *honesty*, *integrity*, *humility*, *patience*, *trust*, *reputation*, and *reliability*. He cites Benjamin Franklin's autobiography as an example. These traits certainly were the core beliefs that our country was founded on. Benjamin Franklin, George Washington, John Hancock, Samuel Adams, and other signers of the Declaration of Independence and authors of our Constitution are widely recognized as possessing and displaying these deep-seated core values. Covey called this the *Character Ethic* and states that it is necessary to possess the traits of the Character Ethic if we are going to achieve long-term success. Following World War I people became more interested in *promoting themselves, advancing in their profession and society*, and *accumulating wealth and material things*. The popular bestselling book *How to Win Friends & Influence People* by Dale Carnegie (Pocket 1998), first published in 1936, is the classic publication on personal success and an example of the *Personality Ethic*. The necessary traits of the Personality Ethic were promotion of self and doing whatever was necessary to climb the corporate ladder or get ahead. These became the accepted norm for the time and to some extent continue today. Instant gratification, accumulation of wealth, and material things became more important than the Golden Rule or high character ethics. Covey calls the last 50 years, the era of the *Personal Ethic*. He states that some traits of this ethic include positive mental attitude, shallow interests in others, and even deception and manipulation of others for one's own benefit. He believes that the Personality Ethic is only a quick fix or band-aid, and it can't be this type of beliefs that lead to our long-term success. Traits such as those of the Character Ethic are important in changing our paradigms.

MASLOW'S HIERARCHY OF NEEDS

Noted psychologist Abraham Maslow published, "A Theory in Human Motivation," in the journal *Psychological Review* in 1943. In it he developed the theory that man had five levels of needs, each level of which had to be satisfied before a person could move up to the next level. Maslow further divided his needs into two categories, deficiency needs and being needs. Deficiency needs began with the lowest level, very basic physiological needs for food, water, and shelter. The next level is safety needs for security of employment, secure revenue and resources, physical security and safety, and moral and physiological needs. The third level of needs is love and belonging. These are basic needs of the human being for friendship, sexual intimacy, and having a family. The fourth and final level of deficiency needs are status or esteem needs. They include respect for self and others, recognition, and lack of inferiority complex. If these deficiency or fundamental needs can't be satisfied, then a person can't be motivated to move to the fifth level, or being needs. This level of needs is broken down into self-actualization and self-transcendence needs. A person who meets the self-actualization level embraces the following:

- The facts and reality of the world rather than denying them
- Spontaneous ideas and actions
- Being creative
- Interest in problem solving, often the problems of others
- Feelings of closeness to other people and an appreciation of life
- Having a system of morality
- Judging others objectively and without prejudice

Maslow summed this up as reaching your fullest potential.

Maslow's Hierarchy is often graphically shown as a pyramid with the fundamental or deficiency needs shown as the broad base at the bottom and the highest level, or being, needs, self-fulfillment at the top (see Figure 2.4). This signifies that all people, in order to effect their beliefs and actions, must have satisfied the lowest level of needs. People can't seek employment, become educated, or have relationships unless they have provided themselves with food, water, and shelter. Primitive man was almost completely consumed by the survival needs. Today, satisfying this need is necessary for the homeless and unemployed if they are to become productive members of society. The next level of needs, safety and security are almost as important as the fundamental level. Everyone needs protection from harm and safety from threats in order to allow them to become educated, obtain a job, raise a family, and become productive members of society. Adults and children who are victims of violence of any kind are examples of not satisfying this need. Women and children who are victims of domestic violence often require considerable counseling in order to regain self-esteem, break the cycle of violence, and reenter society as confident and contributing individuals. Our current conflict in Iraq may also be an example of young males in a society who have not satisfied the need for safety and security and, therefore, have turned to violence to express this need. Most employers assume that their employees have satisfied these two basic needs in order to be productive workers even at entry-level positions. In order for us to be satisfied, motivated, and productive, and even aspire to the next level in our organization, it generally is assumed that belongingness needs are, at least partly, satisfied. If a person

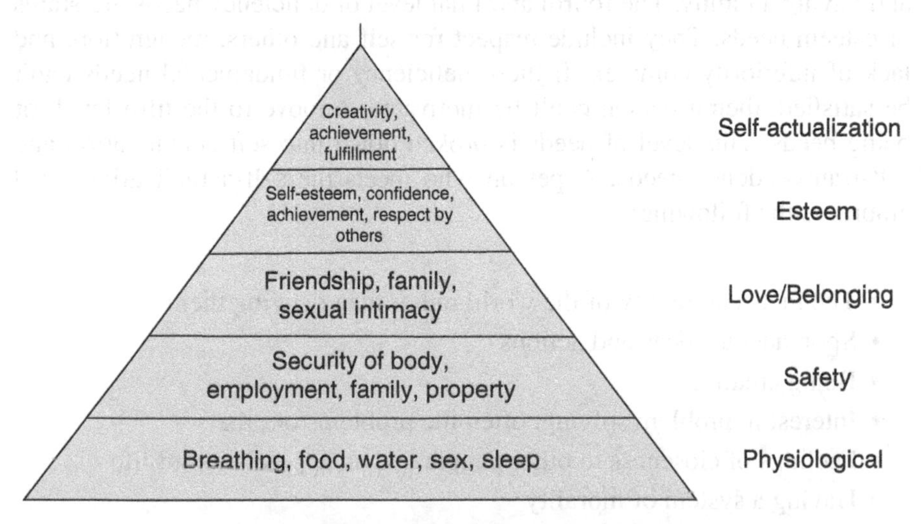

Figure 2.4 Maslow's Hierarchy of Needs.

has an active circle of friends, a loving family or relationship, and a special companion, generally he/she is a productive and motivated employee. The opposite of the belongingness need is often exemplified by an employee who is not productive because they are grieving from the loss of a loved one or going through a particularly stressful period in their lives such as divorce.

The pyramid becomes narrower as we move upward through Maslow's Hierarchy. This signifies that all people do not achieve all five levels of Maslow's needs. The need for self-esteem, respect, independence, freedom, competence, and mastery may only be achieved by supervisors, entrepreneurs, college professors, elite athletes, concert musicians, and others who have achieved a high level of success in their field. The highest level, the need for self-fulfillment, which is displayed at the top of the pyramid, signifies that few of us obtain this level of need fulfillment. Examples of those who do achieve this level are company CEOs, Olympic medal-winning athletes, and recording stars who sell a million albums.

Maslow studied only people with healthy minds during his study. Therefore, successively satisfying his hierarchy is not true for all people. The classic example is the "starving artist." Such a person can be completely self-fulfilled and believe they have achieved the pinnacle of personal success but still not have satisfied lower needs for safety, security, and love and belonging relationships.

THEORY X & Y

Another factor affecting our paradigms and a critical factor in shaping personal change is how we perceive the behavior of others as well as ourselves. MIT professor, Douglas McGregor wrote, *The Human Side of Enterprise*, in 1960 where he described the now well-known Theory X and Y, which describes two very different beliefs regarding man's behavior and attitude toward work and responsibility.

Theory X

McGregor's Theory X centered on the belief that man was inherently lazy and had to be threatened with punishment in order to obtain satisfactory performance toward achieving the goals and objectives of an organization. The specifics of Theory X are:

1. The average human has an inherent dislike of work and will avoid it if he can.

2. People must be coerced, controlled, directed, and threatened with punishment to get them to put forth effort toward achieving organizational goals and objectives.
3. The average person prefers to be directed and wishes to avoid responsibility, has little ambition, and seeks security above all else.

We can think of some examples of Theory X. Many parents believe their children are classic examples of this theory when they are trying to pull them away from the TV or video game and get them to do their homework or something else that the parents consider productive and useful. The male-dominated construction industry may be one of the last bastions of Theory X in the workplace. You can still find old-time construction superintendents who assume their employees are lazy and need to be directed every moment of the work day in order to achieve the project objective of schedule and budget. This behavior is generally recognized by considerable yelling, coercing, and threatening.

Theory Y

Theory Y is a more enlightened and realistic attitude toward people being essentially self-motivated and seeking satisfaction from responsibility and work. The specifics of Theory Y are:

1. Expenditure of physical and mental effort is as natural as play or rest. The average person does not dislike work.
2. External control and threat of punishment are not the only means of bringing about effort toward organizational goals and objectives. People will exercise self-control and self-direction in the service of objectives to which they are committed.
3. Commitment to objectives is a function of rewards associated with achievement of objectives.
4. The average person learns, under the proper conditions, not only to accept, but also to seek, responsibility.
5. The capacity to exercise a high level of imagination and creativity in the solution of organizational problems is widely, not narrowly, distributed.
6. Under the conditions of modern work life, the intellectual potential of the average person is only being partially utilized.

Many examples of Theory Y exist in today's workplace. Industries like engineering and surveying often organize project teams or field crews with a clear understanding of the project objectives and little understanding, at least at the beginning, of how they are going to be achieved. These firms rely on the problem solving, creativity, and responsibility of individuals to solve the problems and meet the client's objectives.

More on Theory X and Y, as well as Theory Z, in Chapter 4.

TIME MANAGEMENT

Many of us have tried different time management systems to attempt to organize our very busy workday. Sometimes we have success for a short period and then give up because the system itself is too time-consuming and "it just doesn't work for me." Over the years I've tried different systems ranging from "to-do list" pads, Post-it notes, and random pieces of paper, to integrated systems of task lists, calendars, and contacts, an organizer program on my desk top computer, a PDA, and a combination of several of these. Sometimes we have success and sometimes, often during a crisis, these approaches just don't work. Why does this happen and is there a better time management system?

Many of our time management problems evolve around several simple practices:

1. Failures to plan at all, finding it easier to let the crisis of the day determine what we do with our time.
2. Failure to separate the important from the unimportant. Often the unimportant items are at the top of our to-do list and are quicker and easier to do, so they get done, while the important but more difficult tasks get put off.
3. Failure to see the big picture. Do you plan from day to day, a week at a time, or do you look into the future at what may be happening in your life and workplace six months, one year and five years from now?
4. Do you have a good understanding of how long it should take for you or others to accomplish a specific task? Do interruptions and distractions cause you to take much longer than originally planned?
5. Are you a procrastinator? Do you feel as though you do your best work when you have a crisis or deadline?

Set Goals

If you answered yes to any of the preceding questions your time management problems may be primarily related to your inability to set goals, both long-term and short-term before you act. You may say "but I'm not a planner" or

"I like to be flexible and respond to situations as they are encountered." Keep in mind that we all plan to some extent. When you left home for work this morning did you turn left or right at the end of the driveway? Did you plan to stop for gas and a coffee? Were you thinking about your appointments for the day as you drove to the office? What about plans for the evening, the kid's soccer game, an important client meeting, or a romantic dinner with your spouse? We do this planning naturally without even realizing or understanding that we are planning. Admittedly, these may be short-term plans and we may not have planned tomorrow, next week, next month or next year but setting long-term goals as well as short-term ones are the key to better time management and more personal effectiveness.

Think about your long-term goals and write them down. Try keeping them on a three by five note card in a place that makes it convenient to review them frequently and remember to review them. Probably the most important thing to remember about long-term planning is that you shouldn't let the fact that things will change stop you from planning. For years, I planned to retire at age 60 and did everything I could think of to plan for it. Now the plan has changed, and I'll be leaving my firm at 64, if everything works according to the plan. Given the path that has taken me to this point, the plan may change again. The most important thing about long-term planning is not the final result of the plan but the process that takes you through the plan. Things change and your goals will change; adjust and revise them as needed

Gordon Culp and Anne Smith in their book *Managing People (Including Yourself) for Project Success* (Wiley, 1992) suggest the following method for setting personal goals.

First, set your long-term goals since they will govern the short-term. Write them down on a sheet of paper during a personal "brainstorming" session. Try to do this in a quiet place where you won't be distracted or interrupted and allow plenty of time. The office may not be the place. You may want to do this at home or better yet on a personal retreat specifically for planning. What are your goals for the following:

- *Personal goals.* Did you always want to learn to play a musical instrument, to speak a new language or take up a new hobby? Have you always wanted to pursue an advanced degree, take a special trip, or visit friends or relatives more frequently?
- *Your career.* Are you happy in the position you are in? Are you working your way up the corporate ladder? Are you unhappy in your present position and it's time to try something else?
- *Family goals.* Are the kids going to college and how are you going to fund their education? Are you an "empty nester" and are you thinking about buying or building a smaller house? Would you like a change of scenery or a warmer climate?

- *Community goals.* How do you fit in your community? Should you get more involved? Have you been active in coaching kid's sports? What about service to professional associations, religious institutions, or non-profits?
- *Health and exercise.* Is your health what it should be to allow you to do the physical activities you want to do? Do you need to loose a few pounds in order to be able to run, bike, hike, or ski better? Do you eat the right things? Do you get the amount of exercise each week recommended by medical professionals as the minimum necessary to minimize your risk of heart disease, stroke, obesity, and diabetes?
- *Financial goals.* Is your financial house in order? Do you know where the check book is if something happens to your spouse and she pays the bills? Are your investments just a pile of certificates and monthly statements? Do you have a will and is it up to date? Do you have a retirement plan and sufficient resources to carry it out?

Once you've roughed out your long-term goals, you need to get more specific. Take another sheet of paper and write each goal in the middle with a circle around it. List the specific actions needed to obtain the goal. Put a circle around your goal and lines with arrows on the actions toward the main goal. These actions necessary to achieve the main goal are your action items. This process will take some time. I recommend that you not have too many long-term goals to start. It will be easier to develop the action items, and you won't be overwhelmed by the process. Perhaps one significant long-term goal and related action items for each of the five categories will be enough for your first time. Remember, plans do change and you can change yours at any time. Remember, the process is more important than the plan itself. Also, remember to put your long-term goals on a 3 by 5 card and place it somewhere where you'll be able to review them regularly.

Once you've completed your long-term goals, your short-term goals will follow. They are the day-to-day, week-by-week, or month-to-month goals that are required in order to meet your long-term goals. Most daily crises don't really fit into the category of short-term goals. They are more likely short-term action items necessary to realize a short-term goal of completing a particular project on time, obtaining new work from a particular client, or keeping your boss happy so that you can move up the corporate ladder. Keep in mind, *most to-do lists do not contain goals but are action items that must be completed in order to achieve a goal*, usually a short-term goal.

In order to separate the short-term action items and your to-do list, from your goals, and to be sure that you are continuing to plan and not just react to the crisis of the day or week, you must set aside some time to review and prioritize your goals. Review the strategies or action items you've set for your long-term goals every few months or at least once a year. Review short-term action items necessary to achieve short-term goals monthly and weekly. The best way I've found is to review my weekly plan on Friday afternoon or Sunday evening for the upcoming week and to review my daily plan for the next day before I leave the office each evening. Remember to put personal items in each short-term plan. This will help differentiate crisis action items from short-term goals, which have to be completed if you are going to have long-term success.

To-Do Lists

Everyone has a to-do list. This is the first item people use to help increase their effectiveness. Although everyone uses a to-do list, few people use it effectively. Many people keep a running tally of items, tasks, or actions to be completed. Most of us try to complete the items at the top of the list before moving on to the next item. What's wrong with this method? The item at the top of the list may not be the most important item you need to accomplish for your short or long-term goals. Few people take the time to organize and prioritize their to-do list.

What are the keys to an effective to-do list?

1. Us a single sheet of paper or a computer organizer like Microsoft Outlook. Don't have scraps of paper or sticky notes everywhere. If you are away from your desk or out of the office often consider a PDA or cell phone that syncs to your organizer so that if you add to your to-do list while you're away or complete an action you can update your list by syncing your PDA and computer upon your return.
2. Keep your to-do list visible at all times. If it is a paper list, keep it on your desk or at your work space. If it is on your computer, keep it on your desktop but minimize it while you are working in order to lessen distractions.
3. Review your to-do list at the end of each day to establish your list for the next day. Add the items not completed today.
4. Prioritize what must be done today. Remember the "80-20" rule. A relatively small portion, 20 percent, of the items on your to-do list are important and will provide 80 percent of the value for your effort today. Remember most small, quick and unimportant tasks fall into the 80 percent of the items that are not important. Concentrate on the important 20 percent, and you will achieve your goals.
5. Include personal items on your list. They provide balance to your life and help emphasize your short- and long-term personal goals.

6. Delegate anything that can be effectively done by others. This is usually those items in the unimportant 80 percent that still need to be done. This leaves time for you to put forth your best effort on the important 20 percent where the results will be received.
7. Add your to-do list to your calendar. If you are using your computer calendar to schedule your day, add the top-priority items from your to-do list with an allocated time for each task. You may not complete all of the items in the time you allocated or even in the order you hoped to do them, but completing this simple task each day will help you realize where your time is wasted by interruptions and help you become a better planner of your time.
8. Measure the results. Remember the old manager's saying *"what gets measured gets done."* Review your to-do list at the end of each day. Review your short-term goals weekly and monthly and your long-term goals at least a few times a year. Measure and adjust your plans as necessary.

The Maturity Continuum

Another aspect of understanding ourselves includes understanding our place on the maturity continuum. It consists of three parts or places;

- *Dependence*. Early in our life we are completely dependent on others for our being and existence. We depend on our parents to provide food, safety, and shelter. When we enter the working world we depend on our bosses, coworkers, and mentors to help us become founded in our organization and to provide guidance, assignments, and challenging work.
- *Interdependence*. This step along the continuum moves us toward working with others. "We can do it" is a team attitude created by interdependence.
- *Independence*. This is where we become self-reliant. We are responsible for our own actions and even our own assignments and clients. "I can do it." is the attitude of independence.

Steven Covey in *The 7 Habits of Highly Effective People* speaks of the three steps along the maturity continuum as being dependence, independence, and interdependence. He believes that interdependence is the highest level of understanding where we work as a team, understand others, think win/win, and synergize. I believe that the maturity continuum can be applied at a basic level of self understanding and used to improve personal effectiveness.

If you are a dependent person, employee, or team player you depend entirely on those around you for almost everything having to do with your being. This level takes a lot of time, understanding, and mentoring by others in order to move you up to the next level. As an employer, I want the dependent members of my staff to move up and out of the dependency level as fast as possible. Dependent people take a lot of time, often are not as productive, and are not as highly valued as team members as those at a higher level. I differ with Covey on whether independence or interdependence is the next level. I view interdependence not as the highest level of personal attainment but the level above dependency. Admittedly, I'm defining interdependence more narrowly than Covey when I say that interdependent people on a firm's staff have moved up to the point where they can work more independently and are more productive but still require guidance and mentoring of others. In my mind, those who have reached a level of independence have attained the highest level of personal achievement. They are not only the most productive people in an organization, but by working independently they have the ability to call on others and establish very productive and effective teams. In my view, independent people can experience true synergy with their fellow workers and clients

INTERRUPTIONS

A major impact on our personal effectiveness is interruptions. All managers of engineering and surveying firms get interrupted regularly and throughout the day. It's part of the job description. Some interruptions are caused by others and some are self induced. Here are some of the interruptions I've experienced and see others in my office experiencing:

Personal Interruptions

- *Email alert or personal messenger.* If you stop what you are working on and instantly respond to each email alert, you are allowing an interruption to affect your personal effectiveness. Turn off the alert and set aside a specific time to return emails. I usually do it the first thing each morning and near the end of the day. I have banned pop-up messages in our office, since I feel they are particularly obnoxious interruptions, which are hard to ignore while trying to continue your work.
- *Telephone calls.* Similar to emails, answering every phone call that comes to you interrupts your work flow. Set aside one or two times a day to return your calls. I return mine in the morning right after I finish my email and in the middle of the afternoon to provide a break if

necessary. Group your phone calls and get as many in as you can in the allocated time. I have a two-minute egg timer on my desk which I flip during a call. If I flip it two times, I determine if I've completed my business and now am visiting with the caller and if it is time to move on to the next call. If I'm really busy I ask my assistant to send all of my calls to my voice mail. I put a personal message on my voice mail each day, which tells the caller where I am that day and when I'm likely to get back to them.

- *Personal cell phones.* I find cell phone calls in the office to be a terrible distraction, as well as another interruption of the work flow. Since cell phone calls often are personal, (and many have their own personal ring tone), they also affect productivity. Just as in church or meetings, ask everyone in the office to turn off their personal cell phone.
- *Web surfing.* Do you have to keep up with the news, weather, or stock market throughout the day? Do you stop everything in order to try and purchase those tickets to the game or show you've told your son or daughter you'd get? The Web is an extremely important part of our everyday business. On the Web we find agency rules and regulations, manufacturers and suppliers, product and material specifications, meeting dates, addresses and phone numbers, and many other valuable pieces of information. I often wonder how we did business before the World Wide Web. But if you wander off to the sports page or another favorite web site you've created another interruption in your work flow. Some firms use software that tracks personal Web activity. In our firm, we still believe in the honor code.
- *Dealing with information.* Do you stop what you are doing each day when the mail comes in or do you just put it in a pile to be dealt with at another time? Information heaps, unread magazines, unorganized files, and other items that contribute to general clutter in your work space affect your personal effectiveness. Try to deal with a piece of paper only once. If it has been in a pile more than a month, should it be filed or thrown out? Do you stop what you are doing to deal with an item on your desk that you've been looking for and has just caught your eye? A clean organized work space is important to your personal effectiveness.

Interruptions Caused by Others

- *Fellow employees.* Sometimes I joke about having one of those number dispensers that you find in the grocery store deli just outside of my office so that I can announce, "Now serving number five" to those standing in line outside my office. As I mentioned previously, if you're a manager, interruptions are part of the job description. I consider it a valuable part of mentoring younger members of our staff. So how should you handle them in order to be personally effective? I generally have an open door policy and plan to be interrupted throughout my work day by members of our staff who need my assistance. I do this consciously. If I have to a task that takes a concentrated effort or a larger block of time, I put a note on my door and ask people to come back later. I also have my phone calls sent directly to my voice mail during this time.

- *Meet fellow employees at their workplace.* I control interruptions by meeting fellow members of the staff at their work stations. That way I can leave in a timely manner and feel as though I control the content of the meeting.
- *Schedule meetings.* If meeting interruptions are seriously affecting your personal work flow and effectiveness, ask others to schedule their meetings with you through your administrative assistant or by utilizing the meeting invitation function of your computer organizer.
- *Visitors.* Sales reps, contractors, clients, family members, and other unexpected visitors often stop by just to say hello. These interruptions can be terrible wasters of your time. I find them one of the worst, since the visitor automatically assumes that you are going to stop everything and spend sometime with them. Sometimes you have to allow for visitors but don't encourage it.
- *Meetings.* Many regularly scheduled meetings are a waste of time. People have figured this out, and that's why you see attendance drop over a period of time at regularly scheduled meetings. If you have them in your office, be sure they serve a purpose other than to just catch up and socialize. Be sure that meetings have an agenda, and ask for it ahead of time. Remember, he who controls the agenda, controls the meeting.

PROCRASTINATION

> "There is nothing so useless as doing efficiently that which should not be done at all."
>
> —Peter Drucker

What is procrastination? A decision not to do something. A decision not to decide. Failing to act when necessary to achieve a goal or complete a task. Some of the signs of procrastination can be found in our to-do list. Do we have high-priority items that just don't seem to get done? Do we fail to plan because it is easier to do the simple things or react to a crisis? They also can be found within some of the reasons for interruptions previously discussed. Do we spend too much time on the phone because we don't want to attack a complex problem? Do we wander around the office or spend too much time socializing because we don't want to address an issue?

What are some of the common reasons for procrastination?

- *We are overwhelmed by the task.* Sometimes a high-priority item on your to-do list overwhelms you because it is a goal and not a task. It describes the end result, but it needs an action plan. Related tasks must be planned and accomplished first before the high-priority goal can be reached. For example, you may be charged with preparing a proposal for a large project and an important client, but you don't know where to begin. Your to-do list only says "Proposal for the ACME project due November 1st," but you are not even sure that you understand the client's objectives not to mention the problem. I'd use what I call the "eat an elephant approach." How do you eat an elephant? One bite at a time! My first action item toward the goals of completing the ACME proposal would be to "do my homework." I'd call the client for more details, which would help me gain a better understanding of the problem. This may lead to assigning parts of the proposal writing to others. John Smith might be assigned to write the wetland identification part of the proposal and an architectural historian may need to write the history section. Do you see what I mean? If you carve the elephant down to bite-sized pieces, you'll have a big meal in front of you, but you're developing a plan to eat it.
- *The task is unpleasant.* All of us have things that we don't like to do that are part of our job description. Probably the most common one for engineers and surveyors is contacting the client with bad news. Nobody likes to hear that their project is over budget and behind schedule, and work outside of the scope is required to solve the problem. These issues don't resolve themselves, and avoidance of an unpleasant task is not the answer. Your client may be upset by the initial shock, and you may be the subject of a reactionary response, but you'll be better off and eliminate future larger problems if you deal with the unpleasant task now rather than later. Another unpleasant task for engineering and surveying managers is staff personnel issues. Are you performing reviews and providing staff feedback regularly? Are you avoiding dealing with a staff member who isn't performing up to expectations? Again, these issues don't often cure themselves and only get worse with time.

These are the two most common reasons that I've seen for procrastination. What are some of the solutions?

- *Take action.* The single most important thing you can do to eliminate procrastination is to take action. Almost any action toward moving a task or goal forward will work. Set a time to work on the task. Define intermediate action items and complete them. Set intermediate action for complex goals. Physically change locations. When I have a high-priority task or goal to complete I move to a clean desk or the conference room. If I fear interruptions I go home to work on the task.

- *Address your fears and assume success.* Once you've figured out how you are going to "eat the elephant," you've addressed your fears. See yourself as preparing a winning proposal that will bring a long sought client and profitable work into your firm. Dealing with the client or staff issue will clear your mind and allow you to concentrate on other high-priority items. These will set the stage for better client relations and mentoring key staff.
- *Reward yourself*. Once you've completed a goal or task that you've been procrastinating over, reward yourself before going on. A cup of coffee, an ice cream cone, or an afternoon spent fly fishing or playing golf is a great way to set yourself up for the next high-priority task, to help clear your mind and motivate you toward future success with less procrastination.

SO WHAT IS PERSONAL PRODUCTIVITY ANYWAY?

Sally McGee, a productivity consultant, has written a bestseller entitled *Take Back Your Life* (Microsoft Press, 2007). She describes the path to personal productivity as setting meaningful objectives in your life, staying focused on them, and developing strategic next action plans to achieve your meaningful objectives. She defines personal productivity as the following actions:

- Getting more done
- Meeting your objectives more effectively
- Using your resources wisely
- Having work/life balance
- Feeling in control and relaxed
- Delegating more effectively
- Spending more time being proactive not reactive
- Going home each day feeling complete
- Getting more of the right things done
- Focusing more on my objectives
- Spending more time with family

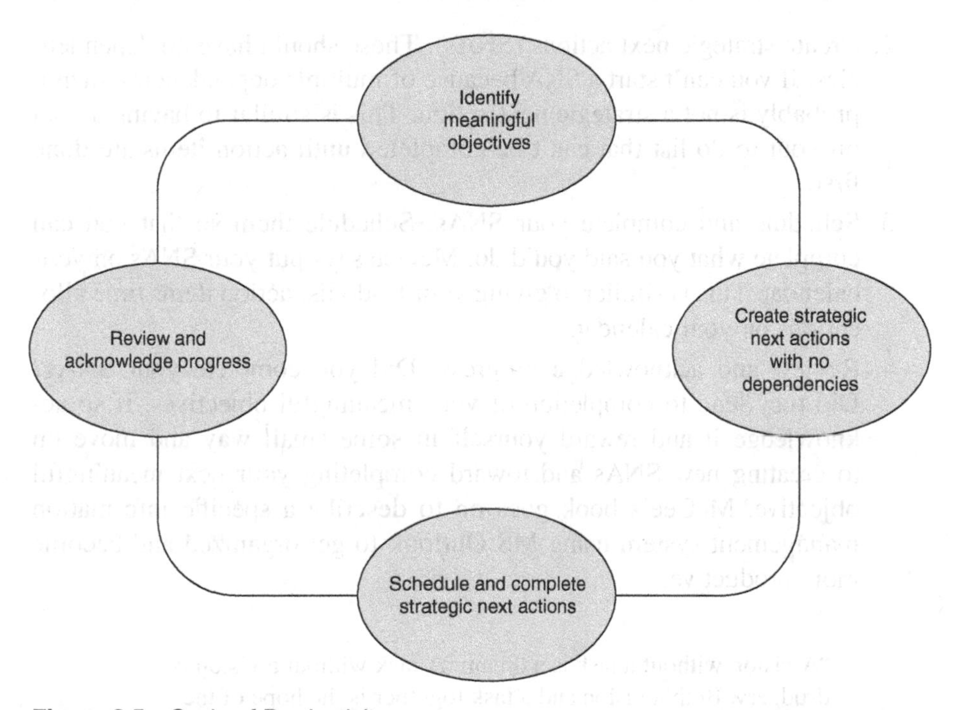

Figure 2.5 Cycle of Productivity.

McGhee, Sally, *Take Back Your Life, Redmond, WA: Microsoft Press, 2007.*

McGee's definition of the first core principle, establish your *meaningful objectives* is similar to the previous discussion regarding determining your long-term goals. She gives four examples of meaningful objectives; sales, career development, health, and home, which are similar to the previously mentioned categories of financial, career, health, and family goals. Whether you call them goals or objectives, most experts agree that if you don't have clear and well-defined goals or objectives, you can fall into the practice of completing tasks or actions on your to-do list with no specific direction or priority just to get them done. Generally, these are the easiest or favorite tasks, not the most important ones. This can result in crisis management, ignoring the important things in life, procrastinating, lack of personal and workplace balance and the feeling of being overwhelmed and life being out of control. McGee states that balance in your life increases productivity not decreases it. She illustrates the cycle of productivity, as shown in Figure 2.5.

McGee explains the cycle of productivity as:

1. Identify your meaningful objectives. This is similar to the exercise described previously to define your long-term goals.

2. Create strategic next actions (SNAs). These should have no dependencies. If you can't start a SNA because of multiple dependencies, then it probably is not a strategic next action. This is similar to having a goal on your to-do list that can't be completed until action items are done first.
3. Schedule and complete your SNAs. Schedule them so that you can complete what you said you'd do. McGee says put your SNAs on your calendar. This is similar to giving your to-do list action items time allocations on your calendar.
4. Review and acknowledge progress. Did you complete your SNAs? Did they lead to completion of your meaningful objectives. If so acknowledge it and reward yourself in some small way and move on to creating new SNAs and toward completing your next meaningful objective. McGee's book goes on to describe a specific information management system using MS Outlook to get organized and become more productive.

> "A vision without a task is a dream: A task without a vision is drudgery. Both a vision and a task together is the hope of the world."
>
> —Sally McGee

SO WHAT'S THE BIG DEAL ABOUT INCREASING PRODUCTIVITY?

A lack of personal productivity leads to, or is caused by:

- Distractions
- Procrastination
- Unclear objectives
- Rework
- Changing and/or conflicting priorities
- Interruptions
- Unexpected crisis
- Technical/equipment problems
- Lack of skill, training or experience
- Misunderstandings
- Burnout/stress/lack of balance

Personal productivity improvements can lead to:

- A higher level of self-satisfaction and accomplishment
- Fewer but more productive work hours
- Balance in your life
- More efficient staff
- Happier staff
- Happier client
- Less need for crisis management
- Increased profits

If you follow the methods outlined in this chapter for self-analysis, invoke change if needed, have clear and meaningful goals and objectives, and use a system of time management that establishes priorities and measures your path toward accomplishing your objectives, you will be well on the way to improving personal productivity and effectiveness.

3

COMMUNICATIONS

"A communication takes place when there is a sender and receiver. Without both there is no communication."

—Unknown

"Unless you are a genius, inextricably linked to the owner of the organization or enjoy other rare privilege, you need effective communications skills to realize your potential."

—Stuart Walesh

We have all heard the old question, "If a tree falls in the forest and no one is there to hear it, does it make a sound?" According to the common definition of communications the answer is emphatically *No!* The tree was the sender of the communication, the sound of it breaking and falling, and possibly a big thump when it hit the ground was the message, but there was no one around to hear it, no receiver, therefore, no communication took place.

Engineers and surveyors are almost universally known as poor communicators, and because of our mostly technical education, we should be. Of the five commonly understood means of communication: listening, speaking, writing, mathematics, and graphics, we spent most of our formal education emphasizing mathematics and graphics. Think of how we chose our technical career. In elementary school, we probably demonstrated an aptitude for mathematics and science. We also likely excelled in geometry and graphic presentation. In high school, we continued our interest with higher-level math and science courses and possibly drafting or pre-engineering. Compositions, essays, spelling, and presentations were our nemeses. Our left brain continued to develop throughout college, while the rest of the world was developing their right brain, their soft skills, including learning how to better communicate.

Some examples of our poor communication skills with clients may include: a poor or nonexistent description of our services and how we will charge for them, our inability to deliver bad news, unwillingness to discuss extra fee when work is requested outside of the original scope of work, and our willingness to "give away the store" in negotiating because we are "nice guys" who want to avoid confrontation. We also may be nervous about making presentations on our client's behalf to regulators. Once we begin the presentation, we may not read the group's body language very well (or present good body language ourselves) and don't know when to stop talking. Although staff may know us better than our clients, many of us have communication issues here, too. How often do we fail to communicate immediately and give feedback to a survey crew member who is consistently late for work, in effect wasting the time of other members of the crew who have been waiting for the person for half an hour or more? How often do we fail to communicate job expectations to young members of our staff who think they are performing well only to face the consequences of a poor performance review later? Do we procrastinate in performing regular performance reviews because we don't want to take the time necessary to analyze a person's performance and then communicate to them how they can improve, in a positive, unemotional, and motivating manner? Do we keep "bad apples" on staff too long while moral deteriorates because we can't bring ourselves to discharge the person for whatever reason? Does it seem as though we always want to give someone one more chance to improve but don't communicate the necessary improvements?

All of the above examples and many more, are examples of areas of communication where we need improvement. In fact, improvement in communication is critical to personal and firm success. This chapter will help you understand the different types of communication and demonstrate how to apply them for your personal and organizational success.

DIFFERENT TYPES OF COMMUNICATIONS

During an ordinary day we *listen* to the kids, spouse, friends, staff, and clients. We respond to their questions, comments, and requests by *speaking* to them directly, having a telephone conversation, or even communicating with them in *writing* by letter, memo, or email. As managers, we probably spend a great deal of time writing proposals and reports. A small part of our day may be spent reviewing *mathematical* calculations or checking *graphics* for a presentation or drawings to be issued under our seal and signature. We probably have not given it any thought, but during a typical day, we've use all of the recognized forms of communication. In order of importance, and our lack of training, they are:

- Listening
- Speaking
- Nonverbal communication
- Writing
- Mathematics
- Graphics

Communications also can be classified as interpersonal or organizational. Each of the above forms of communication takes place within both classes, and we are involved with each everyday. Understanding the different classes of communication is important for personal success.

We also should understand that communications doesn't always work. Sometimes the wrong message is given at the wrong time, to the wrong person.

Since so much of our education was spent studying mathematics and graphics, this chapter will emphasize listening, speaking, nonverbal communication, and writing.

THE COMMUNICATIONS MODEL & MANAGERIAL EFFECTIVENESS

As mentioned at the beginning of the chapter, in order for a communication to take place there must be a sender and receiver. A simple message sent and received demonstrates a successful communication. But there is much more to it. A more complex communication includes the communication channel, noise, and feedback loops as well a predecessor encoding and successor decoding. The communication channel, either written or verbal, is usually the first choice of the sender. This can be based on timing, need, or importance. Once the sender begins the message, it is encoded or translated into easily understood terms or format. It may be a detailed written scope of services describing a project or it may be simple verbal directives to ask an employee to perform a particular task such as prepare a drawing. The message is then decode as the receiver comprehends the message sent through either channel. Noise can affect the receiver's decoding the message. An example of noise may be an interruption during a phone conversation or some other form of distraction that interferes with the message channel. Once the receiver receives the message, it is important to remember that humans have less than perfect memory. Half the message is forgotten within moments of the message being received, and within a few days most of the message may be forgotten. Important messages should use the written channel so that the receiver can continually refer to the message in performing the requested tasks. In order for a communication to be successful, the receiver must have a feedback opportunity. This way they show that the message was received and understood. Additional feedback may be come from the sender to clarify questions or misinterpretations. The communication model in Figure 3.1 is an explanation of the more complex communication that we generally experience.

Managerial effectiveness depends on our ability to deal with complex communications. In fact, some experts say mistakes in judgment may be due to errors in communication as much as 75 percent of the time. A few examples of effective communicators in history who have led successful organizations are;

- Franklin Roosevelt
- Winston Churchill
- Adolf Hitler
- JC Penney
- Bill Gates
- Steve Jobs
- Vince Lombardi

If an organization has effective communications it will be an overall success.

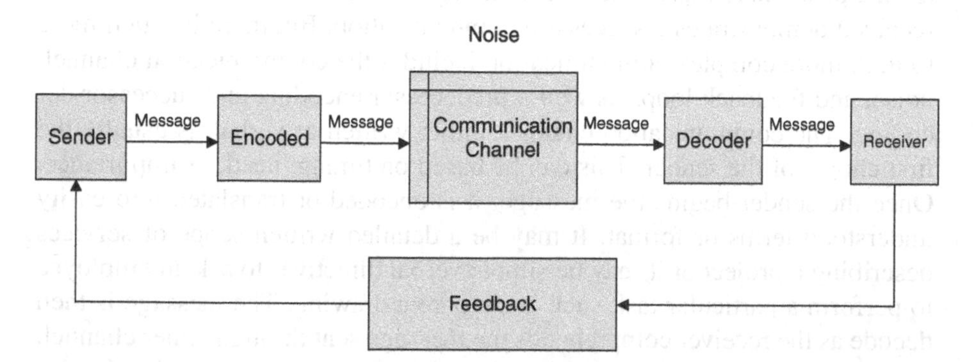

Figure 3.1 Communication Model.

Adapted from Shannon, C.E. and W. Weaver, *The Mathematical Theory of Communication,* 1949.

Encoding—Using the Right Words

Words that we use everyday have many meanings. Variations of the English language and local slang can present communication problems when they are encoded by the sender with one definition and decoded by the receiver with another definition. Examples of a few words that have different meanings within American English and British English are:

- Hood (of a car) versus bonnet
- Trunk (of a car) versus boot
- Windshield versus windscreen
- Vacation versus holiday
- Piton versus peg
- Carabineer versus crab
- Truck versus lorry
- Elevator versus lift

These are just a few examples, and I'm sure everyone can add several more.

LISTENING

Hearing should not be confused with listening. As I grow older, I've begun to loose a certain tonal level of hearing. I seem to have trouble hearing some softspoken people's voices. My spouse often says "Did you hear me?" I'm sure she meant hearing as, our fifth sense, not that I wasn't listening to her. As managers, listening is the single most important attribute we must have.

> "God gave us two ears and one mouth so we can listen twice as much as we speak."
>
> —Unknown

Try this exercise to see the difference between hearing and listening.

Adopt the rule that you can only speak after you restate the ideas and feelings of the previous speaker to their satisfaction. You quickly discover that it is far more difficult than you imagined. You'll also learn that your own views change and emotions drop out of the discussion. Whether you agree or disagree with the other person, being able to accurately experience and restate the other person's point of view establishes real active listening.

> "Listening is that agonizing time I must survive before I can speak."
>
> —Unknown

Why Is Listening So Difficult?

Real listening is very difficult for several reasons:

- We have a natural tendency to immediately judge and evaluate what is being said by another person.
- The average listener comprehends about half of what the speaker says.
  - Within 48 hours retention drops to 25 percent.
  - Within a week retention drops to 10 percent or less.
- Active listening occurs when you avoid the tendency to evaluate and you listen to understand. You see another's point of view and their frame of reference.

Within our ability to listen, retain, and comprehend, there are several levels of attentiveness (Culp & Smith).

- Ignoring—the lowest level
- Pretending to listen
- Selective listening
- Attentive listening
- Empathetic listening—the highest level

Empathetic Listening

Empathetic listening is described in detail in Covey's Habit #5—*Seek First to Understand, Then to Be Understood*. Covey states that empathetic listening is not sympathetic listening. By listening with empathy, you are seeking to understand not to become sympathetic with the sender. In order to listen empathetically, don't push the sender; be patient and respectful. Sometimes silence on our part is necessary to bring out the best communication from others. Covey states that there are four stages of development in empathetic listening:

- Mimic content
- Rephrase content
- Reflect feeling
- Rephrase and reflect

Techniques for empathetic listening include:

- *Silence*. Don't say anything. Let the sender fill the void left by your silence.

- *Comfortable eye contact.* Look at the sender for about five seconds, then look at something else. Don't stare. Intermittent eye contact relaxes the sender and shows your concern.
- *Good body language.* Use body positions that show you are concerned and trying to understand the sender's message. (More on body language later.)
- *Summarize.* At key points in the discussion, summarize your understanding up to that point. Say, "Is this what you mean?" or "My understanding up to this point is. . . ."
- *Signal.* Show that you get it. Tell me more?
- *Reflect your new understanding.* Describe to the sender what you understand the discussion to have been.

Covey says empathetic listening is more difficult than you can imagine. He describes our automatic responses such as evaluating, probing, advising, and interpreting as those that we tend to make if we are not listening with intent to truly understand. Try not to use automatic responses.

The results of empathetic listening are that your own view may change or at least you understand the sender's point of view. Emotion drops out of the discussion and rational debate may take place. Differences are reduced, and whether you agree or not, being able to accurately restate another person's point of view is establishing real communication. Communication can now move toward solving the problem not attacking the person.

Open-Ended versus Closed Questions

> "I kept six honest and serving men. They taught me all I know.
> Their names are What, Why, When and How, Where and Who."
>
> —Rudyard Kipling

Close-ended questions can be answered with a "yes" or "no." They solicit little discussion between the sender and receiver and often are not meant to. Examples of closed ended questions are:

- Are you on schedule?
- Do you think we should do this differently?
- Do you agree with me or not?
- Private, are you standing at attention?

Open-ended questions are intended to inspire discussion, open up communication, and seek alternatives. For example:

- What are the factors that are having the most impact on your schedule?
- What is the way you think we should handle this situation?
- How do you feel about this approach to the problem?
- Private, would you demonstrate the proper way to salute an officer?

As you can see, open-ended questions do not allow a "yes" or "no" answer. They are intended to solicit and engage the other person with the intent of establishing better communication.

SPEAKING

Spoken communication is what we do everyday, often without much thought and sometimes without much skill. Outlined below are the common forms of spoken communication used by engineers and surveyors every day. Some are as simple as an informal greeting of a staff member at the coffee machine. Formal communications and presentations can mean the difference between success in obtaining permits for a client's project or failure to impress a selection committee and not obtain an important job.

Informal Communication

Once we've finished listening, it may be time to speak. This is the area where some engineers and surveyors may have shortcomings. They are not sure when a person has finished speaking or may interrupt another in their haste to "get a word in edgewise." How many of us answer another person's question before they've even finished stating it? The primary type of spoken communication is informal conversations with friends and family. These are interpersonal communications that take place in our everyday work and play. They may take place with little forethought, may be reactionary, and sometimes utilize poor judgment and selection of words. Rarely do they take place with empathy. The consequences here could be being perceived as judgmental, inconsiderate, self-centered, condescending, or interrupting. They could have serious consequences and affect relationships with family and friends or they could be, "just Joe putting his foot in his mouth again." Other informal spoken communication includes telephone conversations, water cooler talk, discussions regarding sports, debate regarding the daily news, and the "office grapevine."

The Office Grapevine

Why is the grapevine so important? It is often the way to learn things within the workplace. The graphic in Figure 3.2 (Boone & Kurtz, 1992) shows that it is where employees get 66 percent of their information, almost 20 percent more than they get from their immediate supervisor.

Often supervisors and management have little knowledge of it. Why is the grapevine so effective? Probably the most important thing is that it is fast. There is no need for messages to travel through formal channels or layers of communication. A message on the grapevine seems to be information that many people desire, find interesting, or need to know, and it is known almost immediately. It also may be selective, utilize discrete channels, and protect confidential information. How often does the boss say "Am I the last to know?"

How should managers interact with the grapevine? Should they try to squelch it? How does it fit with an organization's formal communications? Since many managers have little knowledge of items on the office grapevine, it would be hard to squelch it. The grapevine actually is believed to supplement an organization's formal communication. Some organizations use intentional "leaks" through the grapevine to get important information into

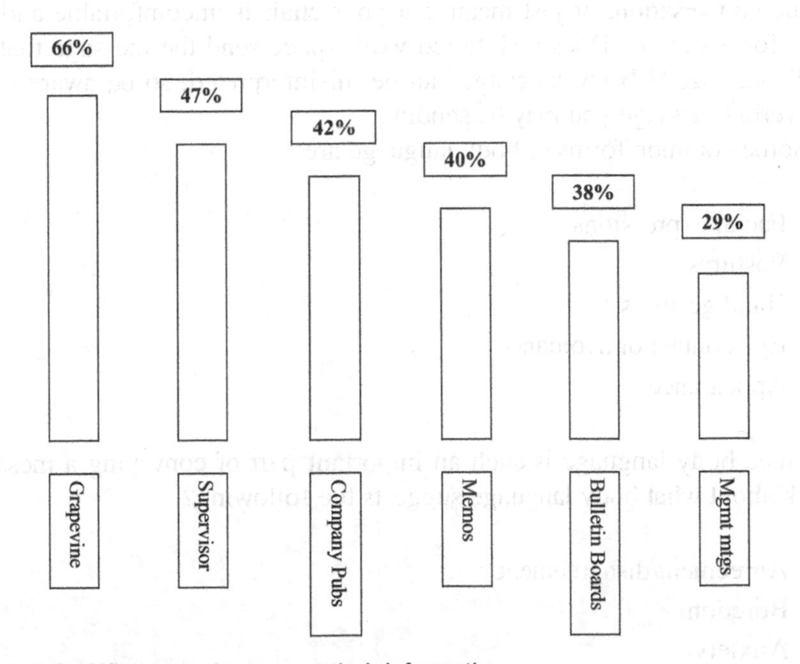

Figure 3.2 Where employees get their information.

Adapted from Boone, Louise E. and David L. Kurtz, *Management* fourth edition. New York: McGraw-Hill, Inc. 1992.

the proper channel, while it appears to be unauthorized. The grapevine is generally considered to supplement formal communications. Therefore, management should seek to understand it rather than destroy it.

Body Language

Body language is another supplemental form of communications. Whether we know it or not, we all use body language. Although unspoken, it can send powerful messages. It helps amplify and clarify the spoken word. Vocal characteristics and facial expressions alone arc said to make up about 50–60 percent of any spoken communication. How do you feel when you enter a colleague's office and he's on the phone and signals you to come in and wait? Body language says the call isn't confidential, and he doesn't mind if you overhear the conversation. What if he signals "stop" with his hand and then a single finger indicating "one minute"? A different nonverbal message is sent. He indicates that the conversation may be something you shouldn't hear, and he'd appreciate it if you come back in a minute or so. What about other body gestures? Does a person rolling up her sleeves mean she is ready to get to work? Does leaning back in one's chair with your hands outstretched over your head indicate your attempt to dominate the discussion, show disinterest in the conversation, or just mean that your chair is uncomfortable and it is time for a stretch? Does a cluttered work space send the message that one is disorganized? Body language can be misinterpreted, so be aware of the nonverbal message you may be sending.

Some common forms of body language are:

- Facial expressions
- Postures
- Hand gestures
- Eye contact or avoidance
- Appearance

Since body language is such an important part of conveying a message, think about what body language suggests the following?

- Agreement/disagreement
- Boredom
- Anxiety
- Confusion
- Willingness to learn more

Formal Communication

Formal or organizational communication takes place every day, mostly at work. As managers, we often are required to resolve conflicts or address uncomfortable situations. Staff performance reviews, hiring interviews, project meetings, and mentoring meetings or discussions also are formal communications. Important formal, spoken communications are contract negotiations, client, committee or public presentations, speeches or lectures, and meetings with regulatory agencies.

I differentiate between informal and formal spoken communication by thinking of informal communication as spontaneous and formal communication as requiring preparation and often rehearsal before making the presentation.

Many engineers and surveyors, especially those who are early in their career, are fearful of, and ill prepared for, formal presentations. The consequences of a poor presentation can be disastrous to the presenter, the client, and the firm. I recently attended a presentation to a client where a young engineer's presentation contained too much technical detail and went far beyond the time allocated. He completely missed the signals of board members who were leaving the room for coffee, falling asleep, and shuffling their papers and closing their notebooks. Fortunately, the only consequence was a few members being late for their next meeting. I mentioned the shortcomings to the young engineer and warned that another client may not be so forgiving.

Presentations as a Formal Communication

All successful presentations start with a good recipe and good ingredients. On weekends, I enjoy making homemade bread. In order to have a successful loaf of deli rye, I need the right combination of water, rye and white flour, molasses, and yeast. Too much of one ingredient, not enough of another, or the wrong baking time and I can end up with a fallen, dense, or burned brick. By following the recipe carefully, I've got a reasonable chance of success. A client or regulatory presentation is the same. The recipe for a good presentation includes:

- *Know the topic*. Be an expert in your area and anticipate questions. Have the answers. Be credible. If you don't know the answer, say that you don't know it but will follow up and find out. If you try to fake it, someone will know, and your credibility will be lost.
- *Know your audience*. Consider their background and knowledge. There's probably nothing worse than presenting unnecessary technical terminology to laypeople. It bores them, is inconsiderate, and your

credibility suffers. On the other hand, if you are teaching a technical seminar or class, be sure that you have a thorough understanding of all of the details and terminology.

- *Know your objective.* Are you trying to increase knowledge, instruct, persuade, or document?
- *Rehearse beforehand.* If you are only given a specific amount of time for the presentation, you must stay within it. Rehearse the entire presentation before a mock audience, with one person timing it. This is necessary to be sure that you obtain your objective within the specified amount of time. Have your graphics organized, and have an assistant run the computer or projector and turn down the lights if necessary. If others are making part of the presentation, make sure that they also stay within the allocated time. Be sure to leave enough time for questions and answers.
- *Making the presentation.* Arrange your material carefully. Have everything organized so that time isn't wasted moving people and things around the room. Emphasize the major points that you are going to make. Start effectively, summarize occasionally and close effectively.
- *Have a plan B.* Many presentations today depend on computers, electronics, and slide shows. Check out the facility and room beforehand to be sure that you have everything you need. If possible, try out the equipment before the presentation. I've seen enough presentations over the years to know that the gadgets don't always work. Don't waste valuable time and fumble. Have a plan B. Back up electronic presentations with transparent overheads, boards, flipcharts, and handouts, if necessary.
- *Pay attention to nonverbal queues and body language of the audience.* Make good eye contact with your audience. Look at someone for no more than five seconds, then move to another. Don't stare at someone, the back wall of the room, or worse at the slide being projected, or read from your notes. Seek good body language and confirmation from your audience. Engage them in the presentation, if appropriate.
- *Dress for success.* Know the local custom for dress. If your audience wears three-piece suits, you should, too. If dress is casual kakis, sport shirt, and tie, you'll likely fit in by dressing the same. A note of caution though—always appear professional, expert, and worth the fee you are asking for your services. Blue jeans and baseball caps are only acceptable for company outings and Saturday in the office. When in doubt about appropriate dress, check with someone who knows. It is better to be overdressed than underdressed.
- *Summarize.* Your finish is the most important part. Restate your objective. Have you answered all of the questions? If you are seeking an approval from a permit agency, board, or commission, ask for it or

restate and clarify the additional information necessary to obtain it at the next presentation. If you are being interviewed for a new project, ask for it. Our firm's presentations always end by stating that we are very excited about the client's project and look forward to having the opportunity to work with them.

- *Seek feedback.* If possible, have someone from your organization attend the presentation to critique it and give feedback. If you know your audience well enough, ask them for feedback after the presentation.

Project Interviews

Over the years, I've been involved in hundreds of presentations. Probably the most important type of presentation you'll ever make is one in which you are seeking work for your firm. It may be you as an individual or more often as a member of a multidisciplinary team. For many years, the qualifications of our firm would get us to the short list interview stage, but we weren't always successful in winning the work. It took years to figure out the formula, but once we did, our success rate improved dramatically.

The strategy for this type of presentation is very important. First, you must do your homework. This consists of knowing your client, the project, and your competition very well. This type of presentation often includes only a few selected firms who have been "short listed" from a field of many. A technical approach to the project is generally also required. This is particularly difficult, since many potential clients know they have a problem but the extent and scope is not well defined. This is where homework comes in. Visit the client and the site before the presentation (sometimes several visits are needed) to gain as much information as you can about the client, their needs, and the project. Federal and state agencies generally do not offer much information about the project before the short list is selected, but once the short list has been chosen, they often hold a briefing or make a staff contact person available to answer questions and provide information to help firms prepare for the presentation. If this doesn't happen, explain to the client that it is in the interest of presenting the best possible solution to their problem to assist you in gathering the information necessary to prepare a first class technical approach and presentation.

Winning presentations leave "no stone unturned" and often include "thinking outside the box." Conventional wisdom and canned solutions that have been applied to similar projects elsewhere generally don't win. When making a presentation, we assume that we've made the short list based on our qualifications and have compiled an excellent team. Therefore, we don't dwell on our qualifications but move right to a discussion of how we will approach the client's problem.

I can't emphasize enough that making the presentation within the allocated time is extremely important. If you make your presentation timely and the potential client goes overtime with the question-and-answer session, consider it a good sign. I also believe that order of presentation is important. If you are first, your presentation must be impressive enough to be remembered through the next presenters. I personally like being the last firm to present. We try very hard to engage the audience right away, and look for body language and questions that indicate that, although the potential client has spent considerable time with those preceding us, they are impressed with our team and our approach to their problem. Also, don't forget to summarize by thanking the client for considering your firm and sincerely ask for the project. You'd be surprised at how many firms don't do this at the conclusion of their presentation.

It used to be that "short list" presentations were made by an experienced and well-rehearsed marketing team or department. Sophisticated clients have become very aware of this slick approach and the potential "bait and switch." They want to see the presentation made by the team who actually do the work. They want to see whom they will be working with everyday. This requires project managers to hone their presentation skills. Their career and their firm's success depend on it.

Meetings

Meetings are a part of everyday life in business. We hold them with the project team to discuss job progress, with individuals or survey crews to discuss assignments, and with clients and consultants to discuss progress in attaining the project objectives. Some people feel that most meetings they attend are a waste of their time. Why does this occur? Generally, because there is no agenda, no specific meeting goals, and no time limit for the meeting. There also may be too much small talk, straying from the topic at hand, and lack of control by the chairperson. People often start a meeting feeling frustrated and leave the meeting with little feeling of accomplishment. This can happen at internal meetings as well as those outside of your organization. You may not be able to control the parameters of meetings that you must attend outside of your firm, but if you are the meeting organizer or attendee of a meeting within your firm, there are several things you can do to make a meeting more productive.

The first and most important thing is to prepare for the meeting. Most meeting organizers have an informal agenda in their mind when they ask others to attend a meeting. If you are the meeting organizer take time to prepare an agenda that details the goal of the meeting, items, or topics to be discussed, who will present various topics for discussion, and how much time is allocated for each topic, or at a minimum for the entire meeting. This not only shows that you are prepared but also that you respect the value of others' time. Start the meeting on time, or if it is the custom within your office to allow a little slack time, start within five minutes of the published start time. This shows respect for those who were timely in arriving and forces those who are late to catch up on their own. It also allows you to finish on time. Remember to practice good listening techniques and display good body language during the meeting. Assign or ask for a volunteer to take minutes and publish them timely following the meeting. If the meeting includes client representatives or others who may not know all of the attendees, pass around a sign-in sheet and request, name, email address, and phone number. You also should take a minute at the beginning for each person to introduce themselves and say why they are at the meeting. If you plan appropriately, this should be the first item on your agenda. At the conclusion of the meeting be sure to summarize the results and review each individual's assignment. If required, set a date for the next meeting. See the sample meeting agenda that follows.

Agenda

Project Management Meeting
9:00 to 10:30 AM
December 1, 2006
Office Conference Room 3

Please respect the valuable time of your fellow staff and show up for this meeting on time.

Item	Person	Time Allowed
1. Update by each PM on deadlines & resources	5 PMs	50 min (10min/ea)
2. Drafting pool schedule	KCB	5 minutes
3. Survey crew scheduling	DCB	10 minutes
4. Marketing update & proposal go/no go	HEB	10 minutes
5. Strategic planning committee update	JJP	10 minutes
6. Other business	All	5 minutes
		90 minutes

Several types of meetings where you can control the agenda are:

- *Preconstruction meetings*. This meeting is held at the beginning of construction. It usually is organized by the engineer or owner. The purpose is to review and discuss specific items related the construction agreement between the owner and the contractor. Items such as schedule, submittals, payment applications, permits, field reports, responsibility for job site safety, change orders, and claims procedures are discussed. Often the "notice to proceed" is issued at the preconstruction meeting.

- *Client/job progress meetings*. These are generally held on a regular basis throughout the project duration. Often the entire project team, or those only dealing with a specific part of the project, meet with the client once a month in order to track job progress, review the schedule, receive updates on financing, and make owner's decisions and changes if necessary. Meetings held early in the project are often to review scope, program, and budget. These are sometimes contentious meetings where everyone is trying to establish their individual "wish list" in order be sure that their needs are addressed by the budget. Inevitably, these meetings result in a project that is beyond the available budget, and the engineer's skill as a moderator, negotiator, and consensus seeker are needed. Sometimes several meetings are necessary before a program and scope are agreed upon. Future job progress meetings address the scope as schematic designs progress through preliminary design and into the construction document phase of the project. Job progress meetings usually continue throughout the construction phase. Detailed documentation is important for all of these meetings. Since project duration is often several years, participants may change and some people tend to forget previous decisions that will impact the work and budget.
- *Internal department or job meetings*. Project teams should meet regularly internally in order to review updated information obtained at client meetings and overall progress through the design phases. These meetings are also used at the start-up of a project to familiarize the entire project team, including consultants, with their individual roles as well as the project schedule and budget. If a project is behind schedule, the performance of a consultant is questioned, or additional resources are needed, the internal progress meeting is where the decisions should be made. Internal meetings are also used to prepare for client or regulatory agency presentations.

Negotiating

Negotiating requires certain mental and speaking skills that many engineers and surveyors initially do not have. Our training prepares us to arrive at technical solutions either individually or as members of a team where there is a common goal and everyone is working in unison toward the goal or solution. Negotiations by their nature generally deal with conflict. You have something I need and vice versa. A position of knowledge or strength provides strategic advantage. The goal of the other side often is to obtain more than they give away, such as more work product for the same or less money. It has been my experience that engineers and surveyors seek to keep their clients happy and don't deal with conflict very well and, therefore, are not good negotiators. They generally are willing to "give away the store," and as a result potential profit, in order to conclude the negotiating phase and get on with the "real work." How can we improve our negotiating skills? As the coach says, "practice, practice, practice." Seek a fair "win-win" outcome and state it up front. Start out with the assumption that engineers and surveyors do business quite differently from the rest of the business world. Many years ago, when I first enrolled in business school for my MBA, I meet business people from many other industries. I quickly learned that the first function of any project in their world was negotiating "the deal." A single "widget" wasn't manufactured or product delivered until the terms of scope, schedule, terms, and finance were discussed, negotiated, and memorialized in writing before any work was actually done.

Many engineers and surveyors are more than willing to either do a substantial amount of work up front, for free, under the title of "marketing," or actually begin work on a project before the entire scope, schedule, and financial arrangements are detailed and agreed upon. I believe this problem is so bad that the consuming public is aware of it and takes advantage of it. For example, a client specifically requests work to be done, by the engineer or surveyor, that they know (or at least suspect) is outside of the original work scope without any discussion of additional fee until the work is done or the engineer or surveyor mentions it many months later. The client has obtained the results of the additional work and is in an excellent position to negotiate that he "felt that the work should have been included in the original scope" or that the "fee be split" between the owner and the engineer or surveyor. In either case, the engineer or surveyor is in a weak negotiating position, which results in lost income and impacts profitability of the project. Another example is the client who asks for additional work outside of the scope, sometimes openly admitting a willingness to pay extra for the work, but the engineer or survey states that he's "pretty sure we can cover it within the initial budget." This obviously is giving away work that could have resulted in additional fee and improved project profitability. In many industries, individuals receive a commission or bonus based on the amount of additional product or service they sell to the customer. This is known as "up selling," and the company is more than willing to share this additional profit with those who generate it.

So how do we deal from a position of strength, negotiate better, and improve overall project profit? The checklist that follows is a good way to start:

— Don't begin any work until you've had at least a preliminary discussion of scope, schedule, and fee budget with the client.

— If the scope is not clear, and you are asked to provide upfront work to better define the project ask to be paid hourly for the work. If necessary, establish a budget of hours and fee that will not be exceeded.

— Provide a detailed preliminary scope of work for your client to review, and obtain their concurrence before determining the fee.

— Provide an honest fee, including a reasonable profit, for your work at the first opportunity. This can be a lump sum fee for various phases or an hourly budget for individual phases of the work.

— If the fee exceeds the client's expectations, negotiate a reduction in scope. Perhaps there is some work that can be done by the client's staff, the number of alternatives to be investigated can be reduced, the number of meetings to be attended can be reduced, or the deliverables can be provided in a digital rather than printed format.

— Memorialize the negotiated agreement in a written form. Most firms have standard agreements. Attach a copy of the agreed-upon scope.

— Be aware of the scope and when work is requested or needed that is outside of the scope, ask for additional fee before the work is begun.

— Negotiate and settle all disputes as quickly as possible. Memories fade on both sides, and things never get better with time.

Expert Testimony

Expert testimony and depositions are one of the most important spoken communications you will ever have as a professional. Both take careful and detailed preparation, and your testimony in both instances will have a significant effect on the outcome of the case in which you are involved. I have been doing forensic investigation and serving as an expert for over 30 years, and I prepare for each case as if it is the first time I've done so. Expert testimony is no place for bravado or a battle of egos with attorneys. This having been said though, sometimes you have to educate an attorney to the difference between his job as an advocate for your mutual client and your job as an expert finder of fact, no matter how the chips may fall. I have often completed an investigation and recommended to the client and attorney that they should attempt to settle the case, since my findings of fact may not be helpful to our client's case.

The most important thing to understand when you are called as an expert witness is that you can be asked for your opinion. A lay witness may only testify as to the facts as they know them. Experts testify on technical matters that are beyond the expertise of the judge and/or jury. Your ability to explain your technical findings in the case is extremely important. You may use illustrations and a report prepared by you to help others understand what you mean. You also may be asked for your opinion regarding the circumstances of what happened or the outcome if another sequence of events had happened.

Depositions are used by the opposing side to try to understand what your testimony will be at trial. The opposing attorney will ask the questions and your client's attorney will be there to object to improper form or leading questions and to cross-examine you in order to clarify answers that he feels can be misconstrued. Since you are representing your client, it is important that you give truthful and factual answers that will be consistent with the testimony you will give when asked the same question at a trial. Depositions are recorded by a court stenographer, so be sure to speak clearly and slowly when answering a question. Don't anticipate the opposing attorney's question and cut him off with your answer. Let him finish the question, be sure that you understand it, then reply. If you don't understand a question, ask for it to be repeated or rephrased. If you think the opposing attorney is trying to back you into answering a question that isn't quite right or that your response could be misconstrued, ask to consult with your attorney before giving the answer. Also answer the question as directly as possible and don't volunteer additional information or explanation. Save that for the trial. Finally, when you receive the transcript of your deposition, read it carefully, correct grammatical or terminology errors, and return the deposition in a timely manner. When the day comes for your expert testimony, reread your deposition and, if possible, prepare with your client's attorney so that you'll know the questions he'll ask and he will know your responses. Remember the opposing attorney's goal, no matter how friendly and complimentary he seems, is to discredit you and your testimony if possible. It has been my experience that one of opposing attorneys' favorite tricks is to ask you the same question that they asked during your deposition and try to solicit a different answer. If they are successful, it immediately destroys your credibility as an expert, and you are no longer believable. Reread the deposition before trial. If you gave direct and truthful answers during the deposition, giving them again during the trial will not be a problem.

I have only touched upon the highlights of a very important spoken communication. If you enjoy expert testimony and make it part of your practice consult one of the texts in the reference section and consider membership in the National Academy of Forensic Engineers.

WRITTEN COMMUNICATION

Most engineering and surveying school programs recognize that the majority of their students may be weak in the area of written communication and are taking steps to improve this important area of practice. Many curricula require written reports similar to those prepared in the real world as part of the learning experience. If a poorly written letter, scope, agreement, or report can be overcome by technically superior performance in the academic setting, the opposite can happen in the real world.

Types of Written Communications

- *Study and report*. Studies and reports of findings are often done as a predecessor to the detailed technical work on a project. They generally are used to communicate alternative solutions studied, establish a schedule and budget for the project, draw conclusions, and make recommendations regarding a preferred alternate or solution. The previous section discussed expert testimony, which generally is preceded by a forensic investigation and report of findings. Studies for small projects may be written by an individual or a team for large projects. Most firms have a style guide, which is used so that all documents published by a firm have a consistent appearance. This is helpful when a report is being prepared by a group of individuals. Project managers generally are assigned the task of editing a report prepared by members of the project team. Their goal is not only for the report to appear in the format and style of the firm but also to edit the writing of individuals so that the report appears to have come from a single source. Some large firms assign this task to a technical report writer/editor, but most firms depend on the ability of their engineers and surveyors to produce a report that meets the client's needs in clear concise language and is consistent with the style and format standards of the firm.
- *Graphics and appendix in reports*. Other forms of written communication are graphics and appendices that are included in reports. The ability to plot full size 22″ × 34″ drawings on an inkjet printer at 11″ × 17″ allows drawings to easily be included in reports as exhibits. Assemble graphics and other exhibits used in appendices with the same care

toward format and style as the rest of the report. Also, be sure to list them in the table of contents and separate each with a title sheet and tab to make locating it easy when reading the report. The recent practice of inserting tables and graphs within the body of a report makes it easier for the reader and minimizes frequent interruptions to go to the appendix.

- *Record and document forms.* Letters, memos, job progress reports, field reports, meeting minutes, and employee reviews are used to document various types of information that are generated regularly and must be memorialized for the record. In our firm, a letter is a formal communication with a client, consultant, vendor, or regulating agency. It typically is used at the beginning of a project to establish the lines of communication or make an official statement for the record. Memos may be used as follow-up to a letter. They also are used to provide additional information or as an informal communication that is not quite as important as a letter. Internal memos with copies to all team members are often used to provide project documentation of important decisions. An important thing to remember about letters and memos is never deliver bad news in a written communication without discussing it with the client first. Put yourself in his/her shoes.
- Field reports document specific conditions found in the field during a field visit. These conditions include the weather, the time of the visit, the workers and equipment on the site, the work being done at the time, the reason for the visit, decisions or clarifications of plans and specifications, field directives to the contractor, and items of concern which may require clarification, testing, or rework. Consistent information in field reports establishes the record of a project's construction and often is used in documenting claims for extra payment. Most firms have a standard form of field report similar to the one that follows.

In our firm meeting minutes are generally issued in the form of a memo. The date and purpose of the meeting is always captioned in the subject line at the top of the memo, since it may be different from the date of the minutes themselves. A list of attendees is always included at the top of the minutes and is used as the distribution list. Important decisions, actions or assignments, including the responsible individual are often italicized or bolded. Minutes and field reports should always include a caption at the bottom that states the reader should review them and immediately respond with any corrections or additions. This is a warning that the minutes or field report is our view of what happened, and if we've missed something or there is disagreement, it should be brought to our attention.

ACME Engineering & Surveying, Inc.

1234 Reality Way
Pleasantville, AX 01234

FIELD REPORT # —
To:

DATE:
JOB NUMBER:
PROJECT:
LOCATION:
OWNER:
CONTRACTOR:
WEATHER

From:

Site Visit

Arrived at site: **Left site at:**

Personnel & Equipment on site:

-
-
-

Work Performed:

-
-
-

Items discussed:

-

Please notify ACME if any information is missing from this field report or has been interpreted differently.

ACME

ACME Engineering &
Surveying, Inc.
1234 Reality Way
Pleasantville, AX 01234

MEMO#1

DATE: August 16, 2005 **ACME Job #: 2003-092**
TO: MIKE SMITH, EMCA CONSTRUCTION
Phone: 603-526-4434 **FAX**: 603-567-8999
FROM: PATRICIA JONES
CC: Matthew Jones, Timber Concepts, Inc, Tony Cousins, & File
SUBJECT: HIGH STREET BRIDGE # 108/099 – BRIDGE MATERIALS

This memo is intended to expedite the shop drawing resubmittal, which was a result of some confusion on the design intent. The memo informs Hansen Construction, the Town, and TCI that the copy of the original shop drawing submittal by TCI was stamped "REVISE AND RESUBMIT." In discussions with TCI, the revisions have been agreed upon, and in order to ensure that the fabrication will occur in a timely manner, ACME is revising our original review to "APPROVED AS NOTED." We anticipate that the following changes will be reflected on the new submittal, which will be furnished to ACME and Hansen concurrent with the initiation of fabrication:

1. A note must be added which states that "All field drilled hole and field cuts must be field treated with Pentachloropheno."
2. The 1 and 1/4″ bolt that attaches the curb to the posts looked to be drawn incorrectly and will be fixed.
3. ACME approves the new deck configuration, with the 3 stiffeners and the elimination of the shiplap joints. The arrangement appears to meet the AASHTO requirements for Longitudinal Glulam Decks. This approval also assumes that TCI will be stamping the final drawings.
4. Likewise, the substitution of single 5/8″ dome head bolts for two 1/2″ carriage bolts is also approved and assumes TCI will be stamping the drawings.

Matt Smith indicated that he anticipates a delivery scheduled for the first week in October. He will confirm this with Mike Hansen by Friday.

If you have any questions or comments, please feel free to contact me.

Please review this memo. It will become part of the record for this project. Please notify ACME immediately of any errors or omissions.

- *Performance reviews.* Like Diogenes with his lamp, I've been searching for years for the perfect employee evaluation form. Several years ago I came to the conclusion that one didn't exist. Most forms have a list of performance criteria and a check-off area ranging from "poor" to "exceeds expectations." They leave little room for comment, and I believe that evaluating an employee's performance deserves more feedback than just checking a box. A memo supplementing the evaluation form has become my standard practice. In it I discuss the employee's performance in detail, utilizing examples from my experience of good and not-so-good performance by the individual. I describe my expectations for the employee's performance, and we discuss his/hers. If we are in agreement, the final outcome is a memo to the employee that states the mutually agreed-upon goals and action plans to be followed up on at the next review. The employee signs a copy, and it is placed in his/her personnel folder. I believe that an employee can easily become overwhelmed by having too many goals, so we try to keep them to a few that are agreed to be attainable. If after several review periods, the goals have not been achieved, you have the documentation needed to discharge the employee if necessary.
- *Scopes and agreements.* The scope of work that you prepare serves several purposes. First, a draft scope is generally prepared as part of a proposal to a client to perform specific tasks in a project for a specific fee. The project understanding describes your concept of the problem to be solved and how your firm proposes to undertake it. Assumptions upon which the scope is based, such as a preliminary study, survey, and geotechnical work to be done by the owner, work not included, and so forth, are also stated up front. Time schedule, members of the project team, basis of fee, deliverables, and number of meetings also are generally included. The purpose is to determine if your understanding of the client's desires is in agreement with their needs. The draft scope forms the basis for the start of negotiations. Usually a scope goes through several iterations before it is finalized and agreed upon by the client. It then is attached to a standard form of agreement that contains detailed "boilerplate" terms and conditions of the contract between the firm and the owner. It is important to use industry standard agreement forms and terms and conditions, since they have been proven in the legal system. Beware of owner-generated terms and conditions that give particular power to the owner, create stringent penalties, and are not appropriate for professional services. Most professional liability insurers will review owner-generated terms and conditions and render an opinion regarding their effect on your insurance. Often you can convince an owner to use your terms and conditions if you have an opinion from your insurer that his are not insurable.

ACME Engineering & Surveying, Inc.

1234 Reality Way
Pleasantville, AX. 01234

Exhibit A

Scope of Services for:
Phase I Environmental Site Assessment
A & C Mechanics
Route 61, Yanbla, NH.
January 22, 2009

The scope of services for the above-mentioned project is as follows:

Assumptions:

1. Work is to be performed in accordance with the scope and limitations of ASTM E1527, Standard Practice for Phase I Environmental Site Assessments.
2. Free access to each site and any associated buildings will be provided.
3. No soil, groundwater, or other sampling or laboratory analysis is included.
4. Inquiry into the presence of lead paint, asbestos, and radon are outside the scope of this assessment.
5. Owner to furnish a copy of the deed and boundary survey of the property.
6. No topographic or boundary survey is proposed.
7. Owner to provide copies of any existing permits related to waste generated on the site including septic system permit.
8. If any wetlands have been impacted, the Owner will provide a copy of the NHDES and Army Corp permits.

Scope of Work:

ACME will provide these services;

1. Make site visit to observe and photograph site conditions and observe the condition of abutting properties that may impact the site.
2. Research available federal, state, and local sources for history and/or presence of potential contamination from petroleum and/or hazardous substances.
3. Conduct interviews with person(s) knowledgeable of the site history.

(*Continued*)

4. Provide a detailed report with a summary of research, site visits, and interviews. Photo documentation will also be provided. A conclusion will be provided regarding the likelihood that contamination does/does not exist on the site. This is not a guarantee or warranty.
5. Provide recommendations regarding proceeding to a Level II – ESA if necessary.

Project Management: H. Edmund Bergeron, P.E., will be the project manager for this work. He will be assisted by Joanne Smith, technician.

Deliverable: ACME will provide two copies of the complete ESA, including Appendices and color photos.

Fee: The lump sum fee for this work is $XXXX plus reimbursable expenses. Travel to the site is included. Reimbursable expenses include copies by others, photographs, express mail, etc.

Schedule: We are prepared to begin work within one week from notice to proceed and complete the work by Friday, February 9, 20XX. Please keep in mind that this schedule is based on prompt delivery of Owner-furnished items specified in the assumptions above and necessary material furnished by others. We can not be responsible for delays that are caused by others. Please keep in mind that reasonable time is needed to conduct thorough and professional work once this information is received.

Agreement: If this proposal meets with your agreement, please sign and return one copy of the attached Letter Agreement.

Letter Agreement

Date: January 2, 2007 Job No: 2007:____

To: Contact
Company
Address
City, State Zip

Phone: (603) 555-5555

Fax: (603) 555-5556

From: Project Manager
Re: Project, location

PM: (Initials)

Dear First Name,

We propose to render professional engineering/surveying services in connection with ________________________________, hereinafter called the

"Project." You are expected to furnish us with full information as to your requirements, including any special or extraordinary consideration for the Project or special services as needed, and to make available all pertinent existing information.

Our scope of services will consist of:

See the General Provisions (Terms and Conditions) on the back of this page for a more detailed description of our and your obligations and responsibilities.

You will pay us for our services on an hourly basis in accordance with fee schedule in effect at the time services are rendered, estimated at $________ to $________, plus reimbursable expenses.

A retainer fee in the amount of - $________ - would be appreciated at the beginning of our work.

We would expect to start our services promptly after receipt of your acceptance of this proposal and to complete our work within ________________. If there are protracted delays for reasons beyond our control, we would expect to renegotiate with you the basis for our compensation in order to take into consideration changes in price indices and payscale applicable to the period when our services are in fact being rendered. This proposal is void after 30 days.

This proposal, the General Provisions, and the fee schedule represent the entire understanding between us in respect to this Project and may only be modified in writing and signed by both of us. If you agree with these arrangements, we would appreciate your signing one copy of this letter in the space provided below and returning it to us along with the requested retainer.

Accepted this ________ day of ________ 2010 Very truly yours,

ACME Engineers, Inc.

By: ________________ By: ________________

Printed Name/Title: ________________ Title: ________________

(*Continued*)

GENERAL PROVISIONS (Terms and Conditions)

ACME Engineering & Surveying, Inc. shall perform the services outlined in this agreement for the stated fee.

Access to Site
Unless otherwise stated, ACME will have access to the site for activities necessary for the performance of the services. ACME will take precautions to minimize damage due to these activities, but has not included in the fee the cost of restoration of any resulting damage.

Fee
The total fee, except when stated as a lump sum, shall be understood to be an estimate, based upon Scope of Services, and shall not be exceeded by more than 10 percent without written approval of the Client. Where the fee arrangement is to be on an hourly basis, the rates shall be in accordance with our latest fee schedule. Reimbursable expenses shall be billed to the Client at actual cost plus 15 percent.

Billings/Payments
Invoices will be submitted monthly for services and are due when rendered. Invoice shall be considered PAST DUE if not paid within 30 days after the invoice date and ACME may, without waiving any claim against the Client and without liability whatsoever to the Client, terminate the performance of the service. Retainers shall be credited on the final invoice. A monthly service charge of 1.5 percent of the unpaid balance (18 percent true annual rate) will be added to PAST DUE accounts. In the event any portion or all of an account remains unpaid 90 days after billing, the Client shall pay all costs of collection, including reasonable attorney's fees.

Indemnifications
The Client agrees, to the fullest extent permitted by law, to indemnify and hold ACME harmless from any damage, liability or cost (including reasonable attorney's fees and costs of defense) to the extent caused by the Client's negligent acts, errors or omissions and those of his or her contractors, subcontractors or consultants or anyone for whom the Client is legally liable, and arising from the project which is the subject of this agreement.

Risk Allocation
In recognition of the relative risks, rewards, and benefits of the project to both the Client and ACME, the risks have been allocated so that the Client agrees that, to the fullest extent permitted by law, ACME's total liability to the Client, for any and all injuries, claims, losses, expenses, damages, or claim expenses arising out of this agreement, from any cause or causes, shall not exceed the

total amount of $50,000 or the amount of ACME's fee (whichever is greater). Such causes include, but are not limited to, ACME's negligence, errors, omissions, strict liability, and breach of contract.

Termination of Services
This agreement may be terminated by the Client or ACME should the other fail to perform his obligations hereunder. In the event of termination, the Client shall pay ACME for all services, rendered to date of termination, all reimbursable expenses, and reimbursable termination expenses.

Ownership of Services
The Client acknowledges ACME's documents as instruments of professional service. Nevertheless, the plans and specifications prepared under this Agreement shall become the property of the Client upon completion of the work and payment in full of all monies due to ACME. The Client shall not reuse or make any modifications to the documents without the prior written authorization of ACME. Documents include, but are not limited to all information transferred to the Client such as CADD files, reproducible drawings, reports, etc. The Client agrees, to the fullest extent permitted by law, to indemnify and hold ACME harmless from any claim, liability, or cost arising or allegedly arising out of any unauthorized reuse or modification of the documents by the Client or any person or entity that acquires or obtains the documents from or through the Client without written authorization of ACME.

Applicable Law
Unless otherwise specified, this agreement shall be governed by the laws of the State of New Hampshire.

Claims & Disputes
Claims, disputes, or other matters arising out of this agreement or the breach thereof shall be subject to and decided by Small Claims Court for amounts up to $5000 and arbitration in accordance with the Construction Industry Arbitration Rules of the American Arbitration Association for all other claims.

Pollution Exclusion
The client understands that some services requested to be performed by ACME may involve uninsurable activities relating to the presence or potential presence of hazardous substances.

Additional Services
Additional services are those services not specifically included in the scope of services stated in the agreement. ACME will notify the Client of any significant change in scope, which will be considered additional services. The Client agrees to pay ACME for any additional services on an hourly basis in accordance with our latest fee schedule.

Scopes of work for small projects often are described in a few sentences within the letter agreement, while ones for larger projects are separate documents attached to the agreement as an exhibit. A typical scope, letter agreement, and terms and conditions was preciously shown.

- *Emails and other electronic communications*. Each of us has been working away on our computer and up pops a notice or ping of an email from a colleague, friend, or client. Often it is a simple request and we minimize the work we are doing and jot off a quick response. Sometime later we vaguely remember what our response was and no record of it was kept for the job file. Hopefully, it wasn't too important. The convenience of email has made it the quickest form of written communication within and outside of our offices, but the convenience and quickness of response often lead to poor or miscommunication, an emotional response, not well-thought-out answers, lack of review by others, and the lack of a permanent record of the communication once the Delete button is pressed. All of this points to the fact that you should not make important decisions or deliver bad news in an email. Since we live in a world of instant gratification, my method of response is to compose a letter or memo as a document file, have it reviewed by another, save a copy in the project folder or other appropriate location and then send it as an email attachment to the requesting party. If an important communication is within the body of the email itself be sure to put a copy in the project file before deleting it.

ROADBLOCKS TO EFFECTIVE COMMUNICATION

Sometimes a communication is not as effective as planned. The sender may have used the wrong channel, the receiver may not have decoded the message properly, or an appropriate feedback mechanism was not established. Some of the common roadblocks (Boone & Kurtz, 1992) that you should be aware of are:

- *Poor timing*. Don't try to communicate with a receiver at the wrong time. When a person is busy, preoccupied with a crisis or continually interrupted is not a good time for an effective communication.
- *Inadequate information*. Provide enough information for the receiver to decode and clearly understand the intent but not too little information to understand the goal of the sender.

- *Inappropriate channel.* Verbal communications are informal and quick but are quickly forgotten and easily misunderstood. Important communications should be delivered in writing. This affords the receiver the opportunity to reread the message and evaluate the sender's intent. Phone calls, like verbal communication, are easily misunderstood and forgotten. I ask employees to keep a telephone log of important telephone conversations. Face-to-face communications allow the sender to emphasize specific points in the communication. If important, they should be followed up with a written memo.
- *Selective perception.* Perception depends on a person's past experience, beliefs, and emotions. A feedback loop is needed to determine if the sender's intent was received and understood.
- *Premature evaluation.* Some receivers evaluate the senders message before the transmission is complete. Empathetic listening helps cure this roadblock.
- *Emotions.* A staff member who is in a lousy mood or having a bad day may not originate or receive a communication effectively. Taking emotions out of a communication is one of the best ways to be sure that it is understood and effective.
- *Beliefs.* People sometimes hear what they want to hear when a communication causes conflict or stress. A feedback loop is needed here to determine if an effective communication took place.
- *Cognitive dissonance.* Here the receiver avoids the message altogether. It is somewhat like not opening your mail or answering the telephone. The receiver often pleads ignorance of the communication. Follow-up no matter what the communication channel may be needed.
- *Rationalizing.* The receiver distorts the message to conform to his/her beliefs or views. Sometimes we rationalize the weather forecast if we plan an outdoor activity and the forecast is questionable.

AVOIDING ROADBLOCKS TO EFFECTIVE COMMUNICATIONS

The sender of a communication can play a role in avoiding roadblocks by taking the following steps:

- *Simplify.* A simple communication leaves little room for ambiguity. Practice "KISS," keep it short and simple.
- *Ensure feedback.* Repeat a question, discuss a complicated communication or quiz the receiver to see if the communication was received correctly. Follow-up, follow-up, follow-up.

- *Improve listening skills*. This is probably the most important step to eliminate roadblocks. Practice empathy.
- *Avoid negativity*. Negative communications lead to miscommunication, emotion, rationalizing, and fear. None of these provide effective communication.
- *Avoid being judgmental*. Judgmental communications often take place when the receiver evaluates before receiving the entire communication. Let the sender finish before you evaluate. Try to put yourself in the sender's position.

This chapter has discussed four of the six forms of communication. Included in each form have been several subsets that are important to understanding communication within your professional practice. An attempt has been to introduce you to each of them, to help you understand them and improve your performance in each of the areas. It also is important to understand roadblocks and how they can derail a communication and how we should try to avoid them. Each form of communication is a topic worthy of a book in itself. If it inspires your interest try reading some of the books in the references to this chapter.

4

LEADERS, MANAGERS, & MOTIVATION

"What is the difference between leaders and managers? Managers do things right. Leaders do the right thing."

—Peter Drucker

"What is motivation? Different strokes for different folks."

—Unknown

Leadership—we all know a great leader when we see one but we are not sure how to define leadership. Vince Lombardi was a great coach and leader of the Green Bay Packers. He took the Packers from a team that won only one game in 1958 to winners of the NFL Championship in 1961. He went on to lead the team to six division titles, five NFL Championships, and two Super Bowl wins. Winston Churchill optimistically led Great Britain with his famous speeches supporting democracy during the German bombings of 1940–41, one of the darkest hours of World War II. He refused Hitler's offer of peace for surrender and convinced of the ability of the Royal Air Force and Navy to stem a German invasion, he sent the British Army to invade Mussolini's Italy, creating frustration for Hitler and another front for Germany to defend. Mother Teresa, also known as the "Saint of the Gutters," had no power over anyone but led projects that supported the lepers and the poor of Calcutta by faith, strong beliefs, conviction, and the example of hard work and commitment until her death in 1997. It is said that she inspired followers, skeptics, and opponents to larger acts of kindness.

How do today's professional sports coaches lead players who are financially independent, represented by agents, and making many times the coach's salary? They have to determine what motivates each player and treat each one as an individual. A common definition of leadership is the use of one's power to influence the behavior of others but leadership is much more than power. There are many kinds of power, some good and some not. They include rewards, coercion, expertise, example, and authority.

You can't sustain leadership without effective motivation. To lead others we must understand what makes individuals tick. What makes engineers and surveyors show up for work every day and put forth their best effort? How do you motivate the survey crew to do field work on a cold, snowy, windy, winter day? How do you get project engineers to perceive and solve difficult technical problems? How do project managers motivate team members to work overtime to meet a deadline? What are people's basic needs? Can we inspire them to a higher level of achievement if they are not met? What are their aspirations? How do you build confidence and trust? How do you address each individual's ego? Does everyone aspire to a position at the top?

In this chapter, we'll review several theories of experts who have studied what it takes to motivate individuals and how they relate to leadership. We'll look at motivation theory from the content or needs point of view and also from the process or behavior change point of view. We'll look at the evolution of leadership theory from the Trait or Great Man Theory to Theory X, Y, and Z and Contingency Theory.

Leaders are commonly differentiated from managers. We see them as driving the boat to achieve success for a client or firm. They have vision and commitment toward goals that others may not even see. Managers are the "bean counters," the keepers of the budget and the bottom line. We expect our project managers to be excellent managers of a client's budget and schedule, but they don't necessarily have to be great leaders. Most classic management texts define a manager as one who plans, organizes, staffs, coordinates, measures, and controls. They carry out the plans and processes of the organization and resolve problems. Managers strive for order and good results. Leaders on the other hand often tolerate chaos and lack of structure. They have vision, are creative, and see the big picture. They may be easily bored once a challenge has been met. In some cases, they may even be narcissistic. To be successful, organizations need both, but are the characteristics of managers and leaders mutually exclusive? Do they have to be two completely different individuals? Even if we have a tendency for one style or another, can we be managers when our organization and clients requires it and leaders when it is necessary to motivate the team or change strategic direction? Modern leadership theory suggests that there are different levels of leadership and management required in each organization and for each job title. Individuals at the lowest level may be required to have a high level of technical skill. They may not be required to lead anyone and only participate as a team member. The next career step is to lead and manage one or two subordinates, while still emphasizing their technical skill. A basic understanding of individual needs and good communication skill may be the only requirement for good leadership at this level. As we progress to a higher level of responsibility, leading and motivating teams and coordinating clients, approval agencies, consultants, and others outside of our direct control, leadership requires less emphasis on technical skill and more on the ability to delegate, coordinate, and understand the needs of others to influence their performance to achieve necessary results and goals. Those at the top level of a firm place even less emphasis on technical skill and more on one's ability to plan strategically and motivate others for high performance. One's ability to be creative, see the big picture, and strategize for the future is also a characteristic of a top leader.

Leadership and motivation are probably not as simple as Drucker's feeling that "managers do things right and leaders do the right thing." This chapter will show how they are integrated in surveying and engineering firms every day and how we must master both as we progress through our career if we and our organizations are going to be successful.

MOTIVATION

Old Theory of Motivation

At one time, it was believed that people were basically lazy and did everything possible to avoid work. This was characterized by Douglas McGregor's Theory X. Bosses motivated workers by coercion, threatening loss of job or pay, and punishment for poor performance with demotion or assignment to menial tasks until productivity and attitude improved. Motivation by intimidation and fear was the common method of employers who were better educated, who were of higher social class, and who held ultimate power over workers. This method prevailed as long as labor was cheap and low-skill workers were needed and plentiful. Later, employers believed that high salaries and generous benefits would attract and retain the brightest and the best workers to their organizations. Today, there is a closer connection between the boss and the employee, and sometimes the distinction between them is blurred. Workers and bosses now work together on projects as part of the same team along with subcontractors and consultants. A higher level of technical expertise may be possessed by the employee and considering today's labor shortages, he is free to move to a new job if he doesn't have a high level of job satisfaction and isn't motivated. Intrinsic factors such as interesting, rewarding, and challenging job assignments; the ability to socialize; and opportunity for advancement are more important than extrinsic rewards such as pay raises and job security.

Modern Theory of Motivation—Content Theories

Today's theories of motivation consider an individual's ability, the work environment, and his/her physical and psychological needs, which affect motivation and performance. Some studies suggest that the best workers are 2–3 times more productive than the worst. What makes these workers top performers? The difference is in style, skill, ability, and attitude. In essence, highly productive workers have ability and are motivated. A common formula for motivation is:

$$P = f(A \times M)$$

In order to motivate a person to achieve maximum performance (*P*), their ability (*A*) is multiplied by their motivation (*M*). It is important to note that both ability and motivation are necessary for high performance. If you have low ability, no matter how hard you try, you will not get high performance. If you have high motivation and some ability, you may get a little further. It has been my experience that people's ability and skill level, can be raised with effective training.

Kurt Lewin, another contemporary behavioral scientist, believed a person's behavior (*B*) was a function of the person's performance (*P*) and their environment (*E*), or:

$$B = f(P, E)$$

Assuming that you've hired a person with ability to learn and the skills necessary for the job, both formulas demonstrate that employers must pay close attention to an individual's needs and the work environment if they are going to motivate the person to maximum performance and retain them.

Initial studies of motivation considered peoples needs as the factors that drive motivation. These are known as content or need theories. In Chapter 2, Abraham Maslow's Hierarchy of Needs was introduced to help you understand that individuals have certain levels of needs, which range from very basic needs like food and shelter to the highest level of self-actualization. To manage yourself to achieve your career goals, and to produce at the maximum level, you must understand that in most situations lower level needs of food, shelter, security, and friendship must be satisfied before you can realize a high level of self-esteem, confidence, and ultimately creativity, self-actualization, and fulfillment. It is important for one to understand Maslow not only in the context of managing oneself but also in terms of motivating others to perform at their highest level. Maslow believed that people perpetually want to perform and achieve a higher level of actualization if certain lower-level needs are satisfied. While Maslow's theory has some problems, it is one of the most widely recognized when it comes to understanding what motivates people.

Frederick Herzberg's theory of motivating and hygiene factors considers job satisfaction and dissatisfaction rather than needs (see Figure 4.1). His test group consisted of professionals, including accountants and engineers. He considers two factors that affect job satisfaction motivators and hygiene factors. Motivating factors such as job achievement, recognition, opportunity for growth, responsibility, and advancement create job satisfaction, while hygiene factors, such as supervision, a chance to socialize, working conditions, monetary rewards and benefits, and job security, are neutral or create job dissatisfaction if poorly provided or not provided. Herzberg's hygiene factors satisfy one's low-level needs, while motivating factors satisfy higher-level needs. Herzberg's finding that people and performance were positively affected by workers having direct client relationships has a specific application in engineering and surveying firms. Providing client interaction for workers early in their career satisfies higher-level needs of self-esteem, self-satisfaction, and ability to influence an outcome. In our firm, as soon as a young engineer or surveyor demonstrates a desire to meet with clients, we give them the opportunity. At first, this is done in conjunction with a senior member of the staff or the project manager. Once the young engineer or surveyor demonstrates that they have good communication and people skills, they generally make all day-to-day communication with the client, reserving only important contract or policy contacts for senior members.

Money as a Motivator

While money was demonstrated by Herzberg to be a hygiene factor, not a motivator, further discussion is needed. Hygiene factors can be neutral or create job dissatisfaction, but alone they cannot create job satisfaction. You cannot convince a dissatisfied employee to stay in a firm by offering the individual more money. You need to provide motivators too. Money is a powerful tool for a manager but not the only tool. Competitive salaries and benefits are paid by all employers and have been determined in several studies to have little long-term affect on employee motivation and retention. If a firm pays competitive salaries, money is likely to be a neutral factor. If pay is perceived to be low, it will produce a dissatisfied worker. In the neutral case, a worker will be motivated to higher performance by more challenging work, a higher level of responsibility, or more recognition for a job well done. Money has symbolic value though. It affects a worker's status and says to others that the worker is more highly valued than those who are paid less. Individual needs

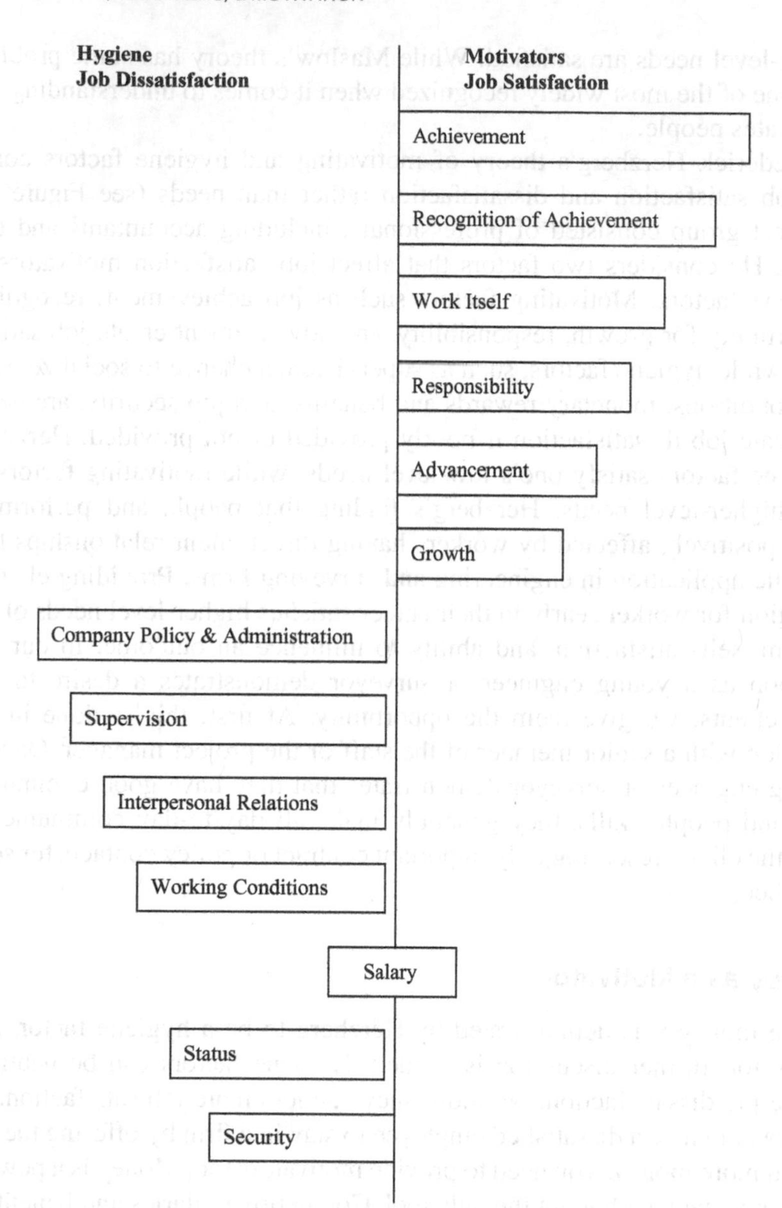

Figure 4.1 Herzberg's profile.

Adapted from Boone, Louis E. and Kurtz, David, L, *Management*, fourth edition, New York: McGraw-Hill, 1992

vary, and a manager must determine if a person's intrinsic needs are being met before expecting that more money will produce a high level of motivation and improved performance. This being said, many low-level needs can be satisfied with money alone.

Intrinsic & Extrinsic Rewards versus Performance

Intrinsic rewards are those produced from within a person, those that produce a feeling of self-satisfaction and achievement. They include most of Herzberg's motivating factors. Extrinsic rewards are those offered by the boss or the organization such as job security, salary, bonuses, and promotions. They include most of Herzberg's hygiene factors.

It is important for managers to understand that not all employees want, or are capable of producing, intrinsic rewards. Some people are perfectly happy seeing their job as a necessary burden that provides the money they need for their style of life. Performance and satisfaction are almost always directly linked by intrinsic rewards and have only fuzzy links to extrinsic rewards such as money, as shown in Figure 4.2.

Modern Theory of Motivation—Process Theories

While content theory concentrates on satisfying individual needs, it provides little direction to a manager for specific action to motivate individuals. Process theory looks at a person's behavior and how it can be changed to be energized, directed, and sustained for high performance.

Operant Conditioning & Behavior Modification Theory

B.F. Skinner developed the Operant Conditioning and Behavior Modification Theory. In it, he stated that there are two types of behavior: reflex behavior, which is involuntary, and operant behavior, which is voluntary. Since operant behavior is voluntary, it is the basis for behavior modification, which can be taught. The goal is to strengthen good behavior and discourage poor behavior. This is done through reinforcements and punishments. Most parents have a basic understanding of this method of behavior modification. A child

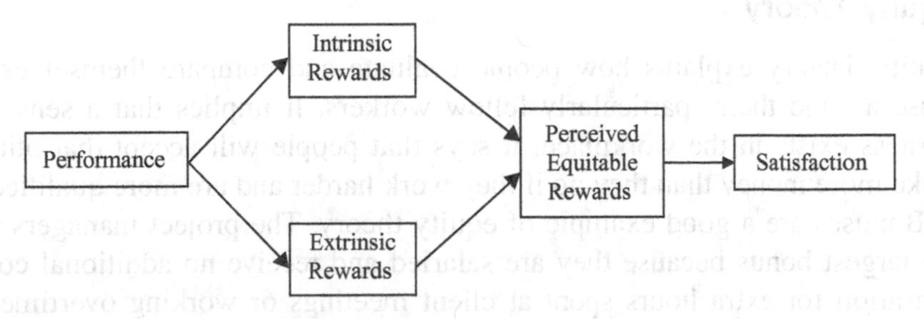

Figure 4.2 Performance–satisfaction graphic.

Adapted from Boone, Louis E. and David, L. Kurtz, *Management*, fourth edition, New York: McGraw-Hill, 1992.

is rewarded (reinforced) for eating his vegetables, doing his homework, getting good grades, and being polite. He is punished or given a "time out" for fighting with a sibling, failing a test, or not completing assigned chores. A baby quickly learns that crying brings reward in terms of food or a bottle. A parent smiling, cuddling, or rocking the baby also brings behavior change through physical reward. Skinner believes that rewards and punishments can be used in the workplace to effect behavior change. Positive reinforcement must be linked to behavior that the organization values. In providing rewards to reinforce desired behavior consider the following:

- Rewards must tie closely to the desired behavior.
- Different levels of desired behavior should receive different rewards.
- Provide rewards frequently and consistently.
- Provide feedback. Workers need to know what they do right and wrong.
- Link consequences. Provide maximum reward for maximum behavior change.

In our firm, we pay bonuses quarterly because we believe that the reward for completing projects successfully and profitably should be closely linked to the timing of bonuses. The project managers have the greatest influence on project success and, therefore, receive the largest part of a project bonus, but the rest of the project team also receives part of the bonus in proportion to their contribution to the project. We've also found that providing constant and consistent feedback improves productivity and performance by minimizing rework.

Critics say that this theory results in only temporary behavior change and that rewards must be continually increased in order to influence desired behavior.

Equity Theory

Equity Theory explains how people evaluate and compare themselves to those around them, particularly fellow workers. It implies that a sense of fairness exists in the workplace. It says that people will accept that others make more money than they do if they work harder and are more qualified.

Bonuses are a good example of equity theory. The project managers get the largest bonus because they are salaried and receive no additional compensation for extra hours spent at client meetings or working overtime to insure project success. Utilizing the Equity Theory, the project manager is the comparison person. An employee who receives lower pay accepts that he is receiving a smaller bonus because he is paid for any overtime hours he works and is not required to attend evening meetings.

Goal Setting Theory

Goal Setting Theory states that people who have or set goals for themselves perform better than those who don't. It also states that people who have more difficult goals perform better than those with easy goals. An employee who spends part of Friday afternoon or Sunday evening planning and setting goals for the coming week will be more successful than those who show up on Monday ready for the crisis of the day or those who are easily distracted from their goals.

In summary, contingent, or needs, theories described lower- and higher-level needs that must be satisfied if an individual is going to be motivated to a high level of performance. It provides the why but not the how to motivate someone. Process theory is commonly believed to provide the means or "how to" for motivation. It says that it is important for people to believe in themselves, and management must provide the expectations and the rewards for their doing a good job. It explains that if intrinsic motivators are provided, people easily rationalize that they are being treated fairly.

LEADERSHIP

Leadership is what the public generally considers management. From their perspective, it is the most visible aspect of running a firm or organization. Many of us in small engineering and surveying firms switch between managing and leading several times a day. Common definitions of leadership include: acts of motivating people to perform tasks to achieve the objectives of the organization, or the act of making things happen. Leaders accomplish this by motivating, directing, guiding, mentoring, and inspiring as opposed to through the pure management tasks of planning organizing, staffing, facilitating, measuring, and controlling. Leadership involves the use of power in ways that inspire and motivate. It is this use of power that allows a leader to invoke behavior change. The five sources of a leader's power are:

1. *Reward power*. Giving employees what they want.
2. *Coercive power*. The opposite of rewards. Threatening or punishment.
3. *Expert power*. Employees respect and follow someone who they consider an expert in a particular field.
4. *Reference power*. An individual who desires to become more like a particular leader uses reference power.
5. *Legitimate power*. The power that is granted by the organization.

Here are some examples: A boss who determines and delivers bonuses is demonstrating reward power. Coercive power may be used when an employee is receiving negative feedback for the third time during a review or whose poor performance is bordering on termination. The boss may have no other choice than to suggest that behavior must change if a person is going to continue in his/her job. Expert power is demonstrated by a technical "guru" in a firm who is the "go to" person when others have problems in a particular area. In our firm, it is the computer systems person, who is called upon when someone's computer doesn't perform as expected. Reference power may not even be realized by a leader who is mentoring a young employee, who wants to become just like his mentor. Legitimate power is granted to those at the top of an organization based on their position and responsibility. It must be understood and used carefully. Often leaders with legitimate power overwhelm or intimidate employees.

Power may be derived from a person's position in a firm or it may be held by someone who has no official standing or position of power in the organization. An informal leader is a person who wields power even though they have no official position of power in the firm. My example of the computer "guru" is an excellent demonstration of informal and expert power. He has no official position of leadership in the firm, but he can have a significant effect on your productivity and related consequences if he thinks someone else's computer problem is more important than yours. I look to informal leaders as future leaders. They demonstrate an orientation toward task accomplishment and have the respect of others in the firm. Senior members or those with a high level of real-world experience often hold reference power over younger members of a firm. These younger members see the road to career success as emulating the senior leader. Legitimate power is authority of position. Those with legitimate power may not be effective leaders if they progress to their position based on a "battlefield" promotion or obtain their position through the "Peter Principle." Formal leaders are people who combine position and personal power.

Similarly to motivation theory, there are three well-known categories of leadership theory.

1. *Trait Theory*. This states that leaders have certain characteristics or traits that can be identified and, therefore, identify a great leader. Early believers in this theory also believed that great leaders possessed these traits at birth.
2. *Behavior Theories*. These talk about different leadership styles. Promoters of different behavior theories believe that leaders can be trained and developed.
3. *Contingency Theory*. This theory holds that no one leadership style is appropriate all of the time but that a leader's style matches the needs of the situation. Success may require an autocratic style in one set of circumstances and a democratic style in another.

Trait Theory

Studies of Trait Theory include the Great Man Theory. It concentrated on who the leader was and the personal characteristics or traits that made him a great leader. Significant political, economic, or social change was brought about by the "great man." Examples include:

- Reigns of various kings and queens of Medieval Europe
- Revolutionary or war heroes like Franklin, Washington, Jefferson and other signers of the Declaration of Independence
- Great athletes, mythical figures, or statesmen, including people like Babe Ruth, the Greek god Zeus, and Daniel Webster

Leadership characteristics of "great men" include: aggressiveness, an undaunting work ethic, and commanding physical features like height, weight, and strength. Doubts in the validity of the Great Man Theory developed when it was realized that some great men possessed characteristics that were diametrically opposite to those of other great men.

Robert Katz, in his 1954 classic *Harvard Business Review* article, "Skills of an effective administrator," pointed out that his article was written to direct attention away from identifying traits of effective leaders and to direct more attention toward identifying the effective skills of a leader. He determined that these skills were technical, human, and conceptual skills. The importance of these skills varied according to where the leader was on the management continuum. Lower-level leaders possess a high level of technical skill, involving the specialized knowledge and experience necessary to do their job.

Foremen or supervisors must have a high level of human skills, since their major function is to work with others. These skills are demonstrated by the way the supervisor perceives his/her superiors, equals, and subordinates, as well as a high awareness of his/her own attitudes and beliefs. Katz also stated that human skills couldn't be a "sometimes thing" and that they must be demonstrated consistently.

The third, conceptual skill is the ability to see the "big picture" or the whole enterprise, including recognizing how various functions integrate into the organization and its place in the industry and community. This skill is required by those at the top level of an organization. Twenty years later Katz wrote a retrospective commentary on his article. He stated that although he still believed developing these skills in leaders is more important than trying to identify traits, his experience in working with business leaders led him to believe that conceptual skills must be developed early in life, possibly as early as adolescence. He now believed that conceptual skills should be viewed as "an innate ability."

The study of Trait Theory was continued and expanded by Edwin Ghiselli, who spent 20 years, up to 1970, studying and testing traits or attributes associated with leadership as an extension of the early studies of the Great Man Theory. He developed Ghiselli's self-destructive inventory test utilizing 64 adjective pairs that either best or least described the attributes of the leader, as shown in Figure 4.3.

	Trait	Importance Value
Supervisory ability	A	100
Occupational achievement	M	76
Intelligence	A	64
Self-actualization	M	63
Self-assurance	P	62
Decisiveness	P	61
Lack of need for security	M	54
Working class affinity	P	47
Initiative	A	34
Lack of need for high financial reward	M	20
Need for power	M	10
Maturity	P	5
Masculinity-femininity	P	0

Importance value 100 = Very important, 0 = Plays no part in managerial talent
A = Ability Trait
M = Motivational Trait
P = Personality Trait

Figure 4.3 Ghiselli's Traits.

Behavioral Theories

Behavioral theories study leadership style, not traits, and the functional interaction of the leader, his subordinates, and the situation.

Ohio State Study

In the early 1940s, Ohio State University conducted a study that initially dealt with military leaders and sought to identify the principal dimensions of a leader's behavior. Two dimensions or attributes emerged from the study. First is the degree of concern that a leader feels for his/her subordinates. It is expressed as friendliness or respect. Second is the initiation of structure or the extent of a leader's task orientation, effort to organize the work effort, instruct subordinates, or clarify the supervisor–subordinate relationship. A high degree of consideration and initiation of structure was found to be the most desirable combination.

Theory X, Theory Y, & Theory Z

In 1960, Douglas McGregor developed his Theories X and Y based on leaders' perception of their subordinates. He discovered that these perceptions may or may not be correct, but people tend to behave according to how they are treated.

A Theory X leader believes that his subordinates are uninspired, lazy, and prefer to avoid work whenever possible. He also believes that they lack ambition and seek the security of a leader-directed environment. Most engineers and surveyors are familiar with the old-time construction foreman or superintendents who motivated laborers by intimidation and fear. They were a classic example of a Theory X leader.

A Theory Y leader believes people find their work rewarding and seek responsibility and productivity. The leader encourages employees to participate in the decision-making process and have open communications with the leader. This is a democratic leader who uses a people-centered approach and tries to create a positive environment where employees can achieve their goals and those of the organization. Engineering and surveying project managers are examples of Theory Y leaders. They need to include their team in decision making in order to motivate them to accomplish the objectives of their clients and the firm.

In 1981, William Ouchi published Theory Z, which proposed that U.S. organizations would benefit from adopting the leadership style and characteristics of Japanese companies whose focus is to involve employees in every phase of corporate life and decision making. His theory also includes providing long-term employment and varied and nonspecialized assignments in order to broaden the employee's experience. This theory was not widely accepted, since the two countries have very different cultures. Imagine involving everyone in the company in all decisions!

Tannenbaum & Schmidt Model

Robert Tannenbaum and Warren Schmidt published their classic article, "How to choose a leadership pattern," in the *Harvard Business Review* in 1958. They proposed that leadership was a continuum, but suggested that it was seven continuous steps from the authoritarian style to the democratic style (see Figure 4.4). They vary from the manager having freedom to make all decisions by himself to managers and subordinates jointly making

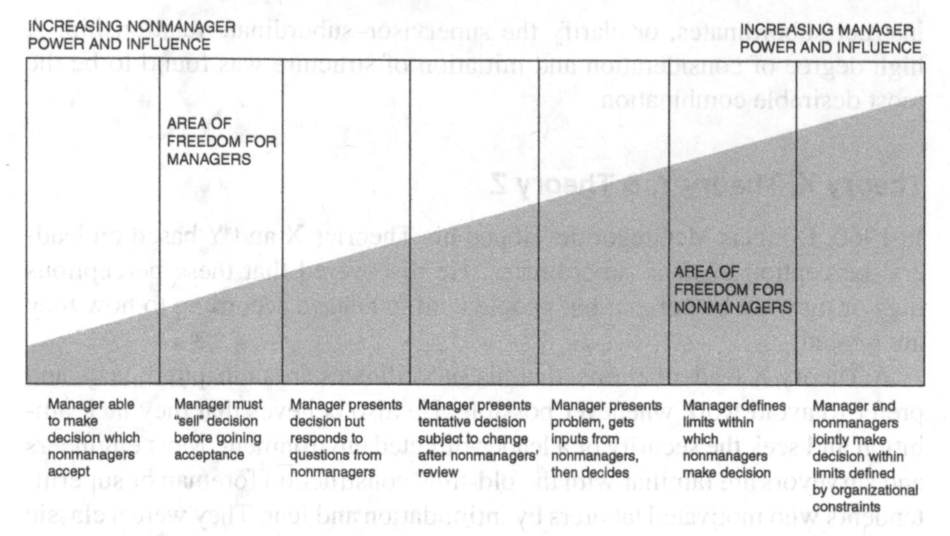

Figure 4.4 The Continuum of Leadership.

Source: Adapted from Robert Tannenbaum & Warren Schmidt, "How to Choose a Leadership Patter," *Harvard Business Review*, May-June 1973.

decisions to accomplish the defined objectives of the organization. Probably their most significant contribution was defining leadership as having three major factors that decide its style and quantifying it in their now widely used formula, which describes leadership style (LS) as a function of:

*The manager (*M*).* His value system and confidence in his subordinates
*The subordinate (*S*).* Their expectations of the managers behavior
The situation (S_1*).* The values and traditions of the organization

$$LS = f(M, S, S_1)$$

In 1973, Tannebaum and Schmidt published a retrospective commentary in which they stated that, although they still believed that leadership was a continuum of seven steps from autocratic to democratic style, the world had become more complex. Outside influences such as the youth revolution, the civil rights movement and ecology and the consumer movement make leadership more of a challenge.

The Managerial Grid

Finally the continuum of behavioral theories was put together in 1991, by Robert Blake and Jane Mouton in their development of the Managerial Grid

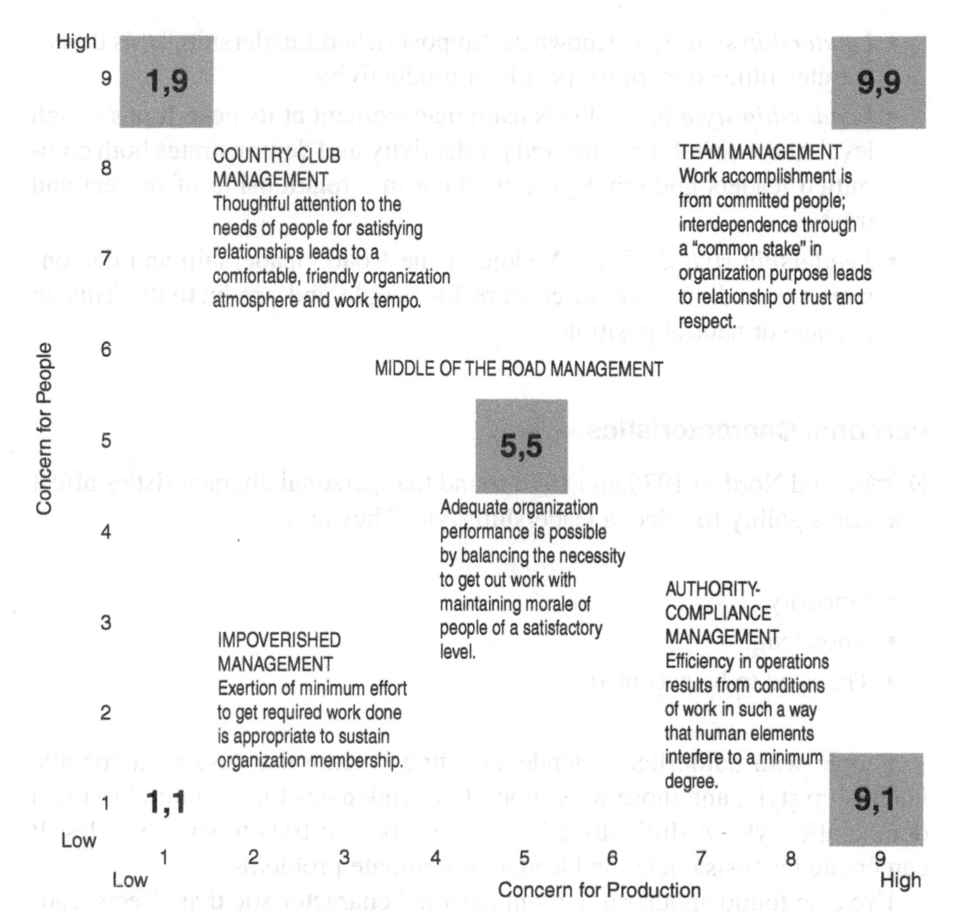
High
9
1,9
9,9
8
COUNTRY CLUB MANAGEMENT
Thoughtful attention to the needs of people for satisfying relationships leads to a comfortable, friendly organization atmosphere and work tempo.
TEAM MANAGEMENT
Work accomplishment is from committed people; interdependence through a "common stake" in organization purpose leads to relationship of trust and respect.
7
6
MIDDLE OF THE ROAD MANAGEMENT
Concern for People
5
5,5
Adequate organization performance is possible by balancing the necessity to get out work with maintaining morale of people of a satisfactory level.
4
3
AUTHORITY-COMPLIANCE MANAGEMENT
Efficiency in operations results from conditions of work in such a way that human elements interfere to a minimum degree.
IMPOVERISHED MANAGEMENT
Exertion of minimum effort to get required work done is appropriate to sustain organization membership.
2
1
1,1
9,1
Low
1 2 3 4 5 6 7 8 9
Low
Concern for Production
High

Figure 4.5 The Managerial Grid.

Adapted from Black, Robert R. and J. Mouton. *The Managerial Grid: The Key to Leadership Excellence* Houston: Gulf Publishing Company, 1964.

(see Figure 4.5). The grid shows behavior as a function of people concerns on the Y-axis and productivity concerns on the X-axis, with each axis rated from 1 to 9. The grid shows five principal leadership styles with many variations in between.

- *Leadership style 9,1*. This reflects authority and concern for high productivity with little concern for people.
- *Leadership style 1, 9*. This is known as "Country Club Leadership" and is the opposite of the previous style. It reflects high concern for people with little emphasis on productivity

- *Leadership style 1, 1*. Known as "Impoverished Leadership," this demonstrates little concern for people or productivity.
- *Leadership style 9, 9*. This is team management at its best. It has a high level of concern for people and productivity and demonstrates both committed leaders and employees working in a relationship of respect and trust.
- *Leadership style 5, 5*. Is "Middle of the Road" leadership and demonstrates a medium level of concern for people and productivity. This an average or neutral position.

Personal Characteristics

Durand and Nord in 1975 and 1976 found that personal characteristics affect a person's ability to select a leadership style. They are:

- Sincerity
- Knowledge
- The need to be accepted

People with authoritarian tendencies find it natural to use an autocratic leadership style, and those with supportive tendencies find it natural to use a democratic style. A difficulty arises when a person tries to switch styles. It can create inconsistencies and leader-subordinate problems.

I've also found another important personal characteristic that affects leadership style is one's ability to follow up. Every leader has more to do than there are hours available in a day to do, hence the need to delegate. Since leaders complete tasks and achieve organizational goals through others, their ability to follow up on progress toward achieving these goals is important to their success. An effective leadership style includes following up on assignments through meetings with project teams or individual project managers. This being said though, the best characteristic a subordinate can have is to keep a leader informed of progress toward completion of tasks before the leader has to follow up. This shows the subordinate is motivated to complete the task and frees the leader to move other projects ahead.

Contingency Theories

Contingency theories demonstrate that effective leadership depends on the situation and that there is no one best way to lead. Due to the complexity of today's firms and organizations, the varying needs of clients and staff, most leaders find that they practice contingency leadership. This style of leadership may be particularly frustrating to those who have autocratic tendencies.

Fiedler's Theory

Fred Fiedler's research published in 1977, led to the concept of Contingency Theory. He developed a test where managers would score their least preferred coworker, LPC. His theory doesn't measure behavior but personality attributes. He found that leaders were relationship motivated or task motivated and that some situations favored one style over the other. Contingency Theory centers on what the manager can and cannot change and then focuses on the solution. It forces the manager to analyze the problem, tasks, and the people involved.

Vroom–Yetton Model

Victor Vroom and Philip Yetton, in their 1973 *Harvard Business Review* article "Leadership and Decision Making," demonstrated that the degree of participation by subordinates affects leadership style. There are five leadership styles and seven questions to be asked in order to determine the appropriate style for the circumstances and the problem to be solved as follows:

Five leadership styles:

1. *Autocratic 1*. The leader makes the decision himself, utilizing the information available at the time.
2. *Autocratic* 2. The leader obtains information from subordinates and then makes the decision.
3. *Consultive 1*. The leader discusses the problem with individual subordinates but makes the decision himself.
4. *Consultive* 2. The leader discusses the problem with a group of subordinates but makes the decision himself.
5. *Group 2*. The leader shares the problem with a group of subordinates. Alternatives are generated and evaluated by the leader and the subordinates and a decision is made by consensus. The leader accepts and implements any group decision but still is ultimately responsible for the results.

There are seven questions to be asked to determine the most appropriate leadership style. These questions require a yes/no answer.

1. Is there a quality requirement that dictates one solution is more appropriate than another?
2. Does the leader have sufficient information to make a high-quality decision?
3. Is the problem structured so the leader can identify the information required, obtain it, and make the decision?
4. Is acceptance by subordinates critical to successful implementation?
5. If the leader makes the decision without consulting subordinates, will it be accepted?
6. Do the subordinates share the organization's goals to be attained in solving the problem?
7. Is conflict among subordinates likely to occur if the preferred alternate is selected?

This model is a typical decision tree where answering yes or no to the questions logically leads to selection of a leadership style. It has been adapted into a computer program, where the leader's problem is entered and after answering the seven questions one or more of the five leadership styles results. I recently tried the program to see if I had selected the right leadership style in attempting to solve a problem in our firm. After answering all of the questions the resulting style was either Consultive 1 or 2. I had elected to discuss the problem with the group involved and, therefore, had selected Consultive 2 as my style. I had made my decision prior to consulting the Vroom-Yetton program and suggest that while it may be helpful to young leaders or those trying to convert to a contingency style, most experience experienced leaders will select the right leadership style intuitively.

Narcissistic Leaders

Freud recognized narcissists as a personality type that seeks the limelight, is charismatic, and has the ability to influence others. Michael Maccoby wrote an article entitled "The Narcissistic Leader" in the *Harvard Business Review* in 2000. In it, he discussed that before the 1980s business leaders shunned the limelight and preferred to run their companies in relative obscurity. During the 1980s, the birth of the Internet and the dot-com business sector saw the rise of narcissistic leaders such as Bill Gates of Microsoft, Andy Gove of Intel, and Steve Jobs of Apple. These leaders were almost universally endorsed as being good for their companies. They had passion and a vision to take the world in a new direction and change the rules of the game. In some cases, they also led their companies to near disaster since they were poor listeners, sensitive to criticism, ruthless, and lacked empathy for others. It is said that the strength of narcissistic leaders is that they "make decisions and get things done." It also is said that their major weakness is "they make decisions and get things done." Narcissists often have a vision that others do not immediately share; it takes time and previous success for their followers to get on board. As a result, they make decisions without consulting others or listening to the advice or warnings of their managers. A strong recommendation is that the narcissistic leader learn to recognize his/her own weaknesses and seek counseling and a leadership style that in more receptive to subordinates' ideas and teamwork. Maccoby has three recommendations for narcissistic leaders to overcome the weaknesses of their personality.

1. *Find a trusted sidekick.* This person will act as an anchor and help ground the narcissist in reality. Most narcissists do not seek close friends, so this person must be sensitive enough to carefully manage the relationship.
2. *Indoctrinate the organization.* Get people and the organization to adopt his vision and think about business the way he does. Jack Welsh of GE said his managers must internalize his vision or leave.
3. *Get into therapy.* This is very difficult to do since narcissists are only interested in themselves and controlling others. Some need psychological therapy to control their rage but are not interested in being self reflective, open, or likable.

Over the years, I have participated in peer reviews of many surveying and engineering firms. I can assure you that narcissists are not limited to the high tech or highly visible industries. It is more likely that there are some narcissistic characteristics in all of us at times.

Emotional Intelligence

Although not specifically a new leadership theory, in 1995, Dan Goleman published *Emotional Intelligence, Why it can matter more than IQ*. He states that IQ is important at entry levels for workers where a high level of technical skill is needed, but as you move on to lead others Emotional Intelligence (EI) becomes more important for success. Throughout his book, he demonstrates that those with Emotional Intelligence possess and/or develop skills of self-awareness, self-regulation, motivation, empathy, and social skills, which help them become successful leaders. What makes up these skills?

Self-awareness. This is a deep understanding of one's emotions, strengths, weaknesses, needs, and drives. He says that we should be honest with ourselves and understand our values and goals.

Self-regulation. This is an inner conversation with ourselves. If you possess EI, you must control your emotions, moods, and impulses.

Motivation. This is a necessary trait of all effective leaders. Goleman says that they are driven beyond expectations. They love to learn and take great pride in a job well done. People with high EI have unfailing energy to do things better.

Empathy. This is the ability to deeply understand others and their make-up. People with high EI seek to see things from the perspective of others.

Social skills. These provide proficiency in managing relationships and building networks. People with high EI are friendly with a purpose and can move others in a direction toward accomplishing goals. They have a wide circle of acquaintances, are adept at managing teams, and are expert persuaders.

Goleman states that those with a high level of EI progress and prosper in leadership, while many with a high IQ flounder.

A Contingency Leadership Example

We had a situation in our firm where a group of subordinates was consistently performing below expectations. Their projects were over budget. They did not have a clear understanding of their assignments, and they consistently failed to meet the client's schedule. This occurred over a period of several months, and it didn't appear to be getting better. In fact, when an inquiry regarding the status of any project was made, no one seemed to know where it was in the process. The project had, in effect, fallen into the proverbial "black hole." I felt it was time to address the situation. First, I consulted with the group leader privately to see if he was aware of the recent performance trend. Upon making him aware that none of the group's projects, which I was associated with, had been completed on time or within budget for the last six months, we decided that a meeting with the group was necessary. At the meeting, I presented the facts for each project, which showed that several projects missed their budget by very large amounts and all missed the client's expected delivery date. This was the first time that the group had seen any data on the projects. Communication of goals and clear expectations was one of the obvious problems. After discussion and input from the group, we agreed that each project had to be "owned" by someone, who would be responsible for tracking it through the process. Someone would be the "go-to" person, and hopefully the "black hole" would be eliminated. In the future, I could determine the status of a project and if it was on schedule and within budget by asking the "go-to" person. It was agreed that the best way to communicate the budget, schedule, tasks, and goals would be a brief project start-up meeting and a checklist of the tasks to be completed, including the project budget and schedule right at the top.

This is an example of my application of the contingent style of leadership utilizing Vroom's Group 2 style, following the decision tree to answer yes to questions regarding quality and acceptance of the decision by subordinates. I didn't start out with the specific goal of applying a certain leadership style or seeking answers to specific questions. My overall goal was, to shape a solution that would be supported by staff and meet the goals of our clients before the firm's reputation suffered any more.

Leadership in the Public Arena

In 1997, Richard Weingardt published a paper in American Society of Civil Engineers', *Journal of Management in Engineering*, entitled, "Leadership: The World Is Run by Those Who Show Up." In his paper, Weingardt speculates that the public sees engineers as those who make things run, and not as those who run things. He reports that the lack of recognition of engineers as leaders is based on poor visibility, the public's perception of engineers, and inadequate involvement in our communities. He sees engineers as having the "right stuff." Contrary to what many engineers think, we possess basic skills that are necessary for leadership in the public arena. Engineers have technical ability, have the motivation to do something that adds value, and are honest. He suggests that engineers and surveyors who have the desire to become leaders outside of our industry take five specific steps:

1. *Get involved in politics*. If you don't have the desire to run for public office, support those who do, then advise them on issues of technical importance of which they may not be aware.
2. *Serve on public boards and commissions*. In many cases, these are appointed positions, which are filled by those who volunteer or just consistently show up at meetings and express their point of view.
3. *Speak out—write and lecture*. In Weingardt's mind this is the most important way to reach the public. Many general interest publications seek to demystify technical projects. If you choose to do this, be sure that your writing skills are good enough so that you can explain in lay terms and omit technical terms. Also many civic organizations seek presentations and discussion on topics of interest to engineers.
4. *Join and be active on issues-driven coalitions*. Engineers have the ability, skills, and background to participate and bring truth over emotion to many issue-driven organizations.
5. *Give back to your profession*. Support and become involved in your profession. Lobby and advocate for the position of engineers and surveyors

at legislative hearings. Remember Teddy Roosevelt's quote, "Every man owes part of his time and money to the business in which he is engaged. No man has the moral right to withhold his support from an organization that is striving to improve conditions within his sphere."

In other words, the world goes to those who show up!

SUMMARY

As you have seen throughout this chapter, leadership and management overlap considerably. In small firms, we don't consciously take off our manager hat and put on a leader hat. Drucker's quote "managers do things right, leaders do the right thing" is ideal at best in the real world. To motivate our staff we must be sincere, honest, compassionate, and understanding of each individual's personal preference and goals. Theories of motivation help us understand that certain needs people have probably have to be met before a person can progress to the next level, but a person's motivation and resulting performance is a complicated issue. We've learned that performance also is a function of employees' ability, environment, and motivation. Some view their job as neutral, neither providing satisfaction nor dissatisfaction. Their job is merely a necessary means to "put food on the table." Many leaders have heard this neutrality expressed by such a subordinate, who says, "I work to live, not live to work." Each theory of motivation discussed has its limitations. Each assumes that leaders select the appropriate style to motivate the individual, but group dynamics can complicate the situation. Certainly, the consensus of those being led is necessary for the long-term success of the leader and the firm. We've seen that leadership theory has evolved considerably over the last 50 years. No longer do we search for those who posses certain "great man" traits, but we seek a leadership style varying from autocratic to participatory and democratic, which can be learned and is applied over a continuum depending on the circumstances and the individuals being led.

Remember, disgruntled employees seldom leave poor organizations; they leave poor leaders!

5

PROJECT MANAGEMENT

"Success and excellence in project management means really managing your projects."

Travers Topgun, LLS, is president and owner of ACME surveying and engineering, a 10-person firm. His firm has more work than it can handle and limited resources. He just received a new assignment from an old and valued client. He's been in business many years and has done this type of project many times before but is concerned that if he doesn't perform this time, his client may switch to a new, local firm. Travers seldom delivers his work product when promised and almost never within budget. He doesn't always clearly understand the client's needs and goals. He has the attitude that based on his experience he knows what the client needs and if the client would just leave him alone he'd get the project done. In addition, he works too many hours while the rest of his staff works 8–5. He realizes that project management training may help him improve his overall efficiency and to deliver his projects on time and within the client's budget, but he just doesn't have time to attend a seminar.

Jock Young, P.E., is a recently licensed young engineer who has become a project manager (PM) after demonstrating that he can work independently. He has excellent technical skills, makes few mistakes, insists on high quality, and has successfully completed several small projects on his own. This was a "battlefield promotion," and Jock hasn't received any formal project management training other than a two-day seminar and reading a book on project management from his boss. Jock is used to working mostly by himself, and in his new position as PM tries to do much of the technical work himself. He's found that in his new roll he spends considerable time reviewing and "red-lining" the work of others. The result is a considerable amount of rework, which he knows he could have done right the first time. He often questions his team's commitment to producing quality work. He is somewhat of an introvert and has trouble communicating with his clients, members of his project team, and others in the firm. He doesn't understand the office PM/financial software reports and has trouble controlling costs and billing work in a timely way. Finally, he has no real power in the organization, must work with other project mangers to secure staff resources, and has poor support from his superiors. They seem to want him to "get to work" and not spend time planning and in meetings. Jock is wondering if project management is really for him.

These examples, a senior principal in a surveying and engineering firm and a young engineer recently promoted to project management, are all too common in many small surveying and engineering firms. In most cases, our technical training does not equip us for project management. In this chapter, you'll discover that project management is more than just kicking butt and living by the numbers. Project management is people management, both client and members of the project team. You'll learn about project management as a complete system, including managing people, processes, and overall project quality. With these skills, you'll become a better project manager!

AN INTRODUCTION TO PROJECT MANAGEMENT AS A SYSTEM

What is a project? It's a problem that needs to be solved. It's created to achieve specific results or an outcome. It's temporary in duration and ideally, it has a start and end date. Projects also have budgets and require the commitment of resources. Projects are what engineering and surveying firms do, and successful projects are key to our success.

What is project management? Simply stated, project management is obtaining, planning, organizing, and controlling projects. But it is not quite that simple. First, we need to clearly *define the project* (the client's problem), not only for ourselves but for our client, in order to assemble the right project team for the work ahead. This includes gaining a clear understanding of the client's goals and objectives, including the correct project name, who the client entity really is, the team likely to be assigned to the work, consultants and other stakeholders, and the desired outcomes for the project. Next is a detailed preliminary project scope including: assumptions, phases and tasks to be performed, schedule, budget, and deliverables. Unless it is a very simple project, I like to negotiate the scope of work before determining the fee. Once the scope is agreed upon, determination of the fee is relatively straightforward. Upon completing the fee negotiation, the agreement is generally signed, a detailed work plan is developed, and work begins. Each of these steps should be communicated to the entire project team. James P. Lewis, in his book *Project Planning, Scheduling & Control* (McGraw-Hill, 2005), suggests this be summarized in a project charter, a document that contains the project purpose, goals, mission statement, brief description of scope, preliminary budget, and lists the project team members and stakeholders. An important part of the charter is the anticipated project outcome or objective. This may be a survey or subdivision plan, an engineering study, or construction drawings and specifications. The team members are required to sign off on the charter to indicate that all agree on the definition of the project to date. Second, is the *planning and strategy phase*, (also known as the basis for the project work plan) in which we plan the entire project from the beginning so that the entire team understands the strategy and overall approach for accomplishing the project's objectives. This includes getting organized to actually do the work. It includes a project organization chart or task and responsibility matrix, detailed schedule and commitment by management of the necessary resources, and delegating tasks to certain members of the project team. Establishing standards and level of quality, as well as defining the final work product and deliverables, is included in this phase. The result of this phase is a detailed scope of work, which forms the basis of the commitment and agreement between the surveyor/engineer and the client. Figure 5.1 shows a flow chart that outlines phases I and II of this project management model.

The *implementation phase* starts with a project kick-off meeting and also includes a project work plan, for complex and long duration projects. It will contain more detailed information than the scope, which is presented to the client. The work plan is the internal document for actually doing the work. It is the detailed plan that dots all of the i's and crosses all of the t's. It ensures that the level of resources committed is available and that the level of quality and standards are clear. Checklists should always be included in the work plan. It attempts to minimize rework and waste based on a lack of understanding and poor communication. Doing it right the first time is emphasized to the entire project team in the work plan. A detailed schedule, including milestones, frequent project meetings, and reviews by the client, permitting agencies, and stakeholders are included in the work plan. The fourth phase is *controlling* the project. This includes managing changes requested by the client and others or those that evolve as the problem becomes clearer. It also includes regular progress meetings with the team and client. Tracking the budget, billing, collections, and changes in scope also are included in this phase. Regular job progress reports to the client also help keep them informed regarding progress of the work during the past month, decisions to be made, and performance according to budget and schedule. (See Figure 5.2).

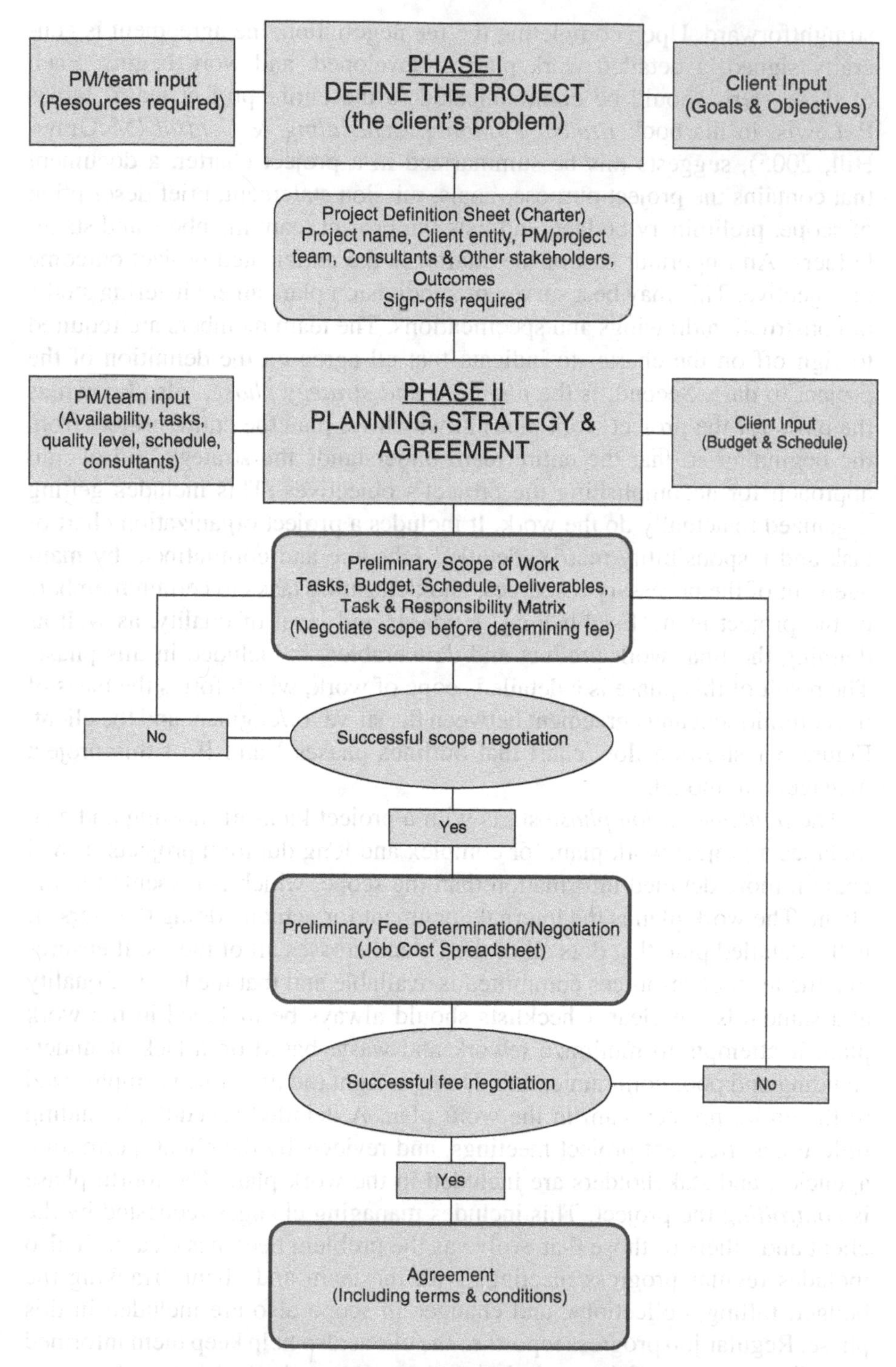

Figure 5.1 Project management phases I and II.

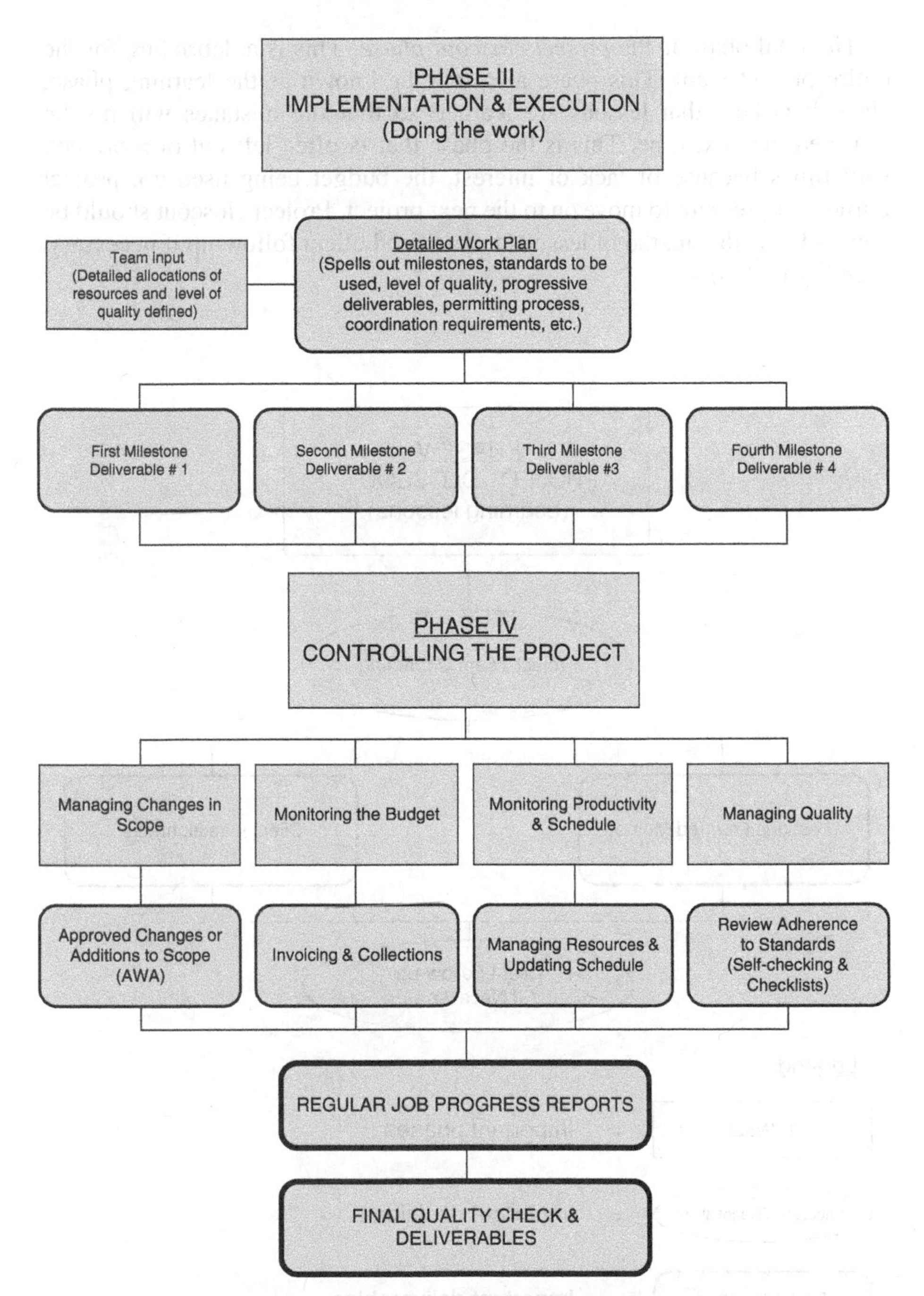

Figure 5.2 Project management phases III and IV.

The final phase is the *project closeout phase*. This is a debriefing for the entire project team. This phase also may be known as the learning phase, since it is here that lessons are learned so that the mistakes will not be repeated the next time. This is the phase that is often left out of a project, sometimes because of lack of interest, the budget being used up, project burnout or pressure to move on to the next project. Project closeout should be concluded with a memo of lessons learned and client follow-up if necessary. (See Figure 5.3).

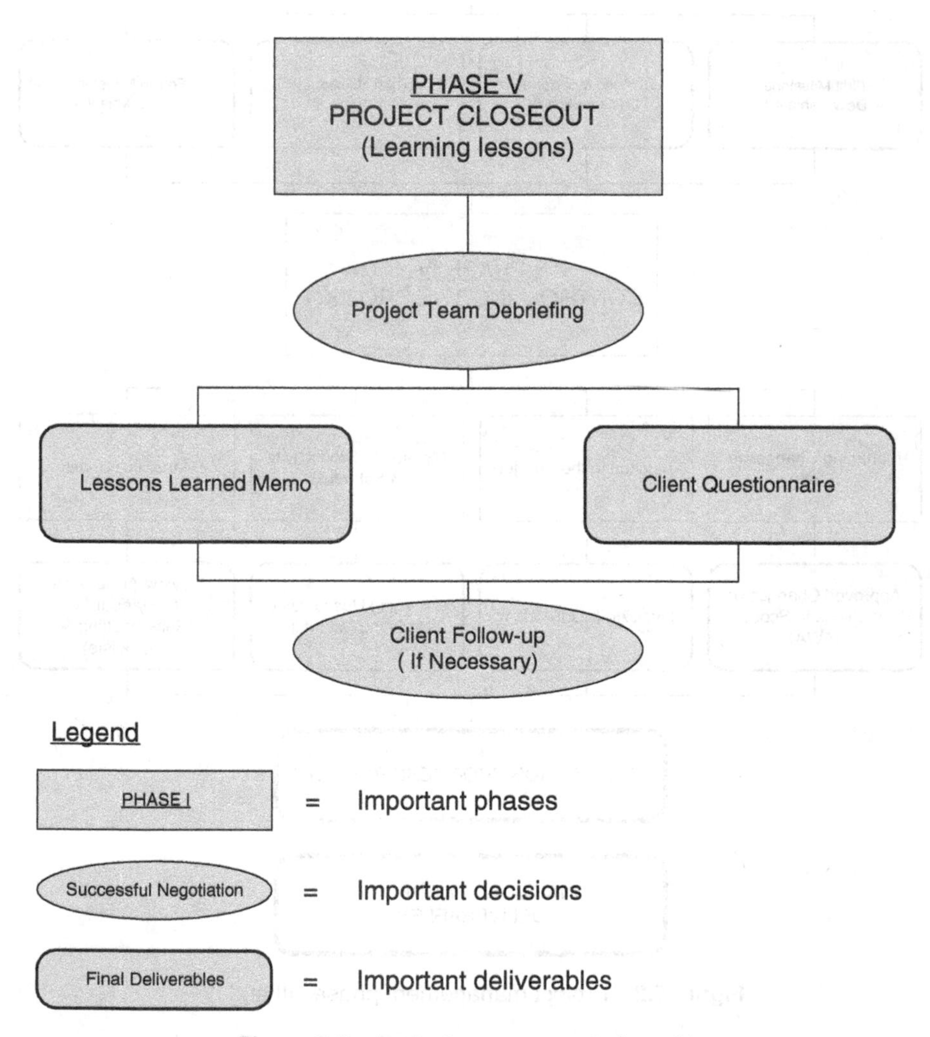

Figure 5.3 Project management phase V.

PITFALLS OF NO PROJECT MANAGEMENT

Lewis states that there are two types of project managers. The first is *dedicated* project managers, who are proactive and take the initiative to be sure that their projects run smoothly. The second is *accidental* project managers, who constantly react to project crises and seldom plan ahead. He also states that a lot of people want to be managers, but many of them don't really want to manage. What are some of the pitfalls of not doing project management or doing it poorly? In the case of both Travers and Jock, their projects probably experience some of the following characteristics.

1. Vague Objectives

Travers feels, "I don't need to plan. I've done this type of project many times before." Since I'm working with limited resources and spend too many hours, the time spent planning is wasted, and we need to just get going and do the work. As a result, the project is not clearly defined, and client expectations and level of quality are not clearly defined or understood. This likely will result in being behind schedule and over budget, again.

2. Vague Responsibilities

Jock Young hasn't received any formal training in communications and delegation so his team suffers from, "Who's on first?" syndrome. The members of the project team are not clear on who is doing what, when, and how. There is a lot of wasted motion and rework due to poor communication and lack of understanding of the client's objectives. Phases, tasks, and responsibilities must be clearly defined and the project manager must be recognized as the leader. He also must lead.

3. Heading down the Wrong Road

Probably Travers field crew started the survey without adequate research and reconnaissance. In essence they are starting work before a clear definition of the problem and with little or no project planning. This results in wasting time, redoing work, or requiring additional trips to the job site to obtain information and detail not obtained on the first trip. It also likely will result in getting behind schedule and being over budget. It has been my experience that surveyors in particular often attempt to solve many related surveying problems that are not included in the original scope and for which their client is unwilling to pay. Sticking to the original scope or communicating necessary changes in scope to the client are important here.

4. Poor Communication

Both Travers and Jock may be guilty of poor communication with the project team and the client, since the team and the objectives were not defined at the beginning and the client wasn't involved in verification of the definition of the project. Scope changes also need to be communicated and agreed upon with the client before the work is begun. If you ask for additional payment based on extra work after it has been completed, you are not in a good bargaining position if the client refuses to pay based on lack of communication and prior approval.

5. Poor Record Keeping and Control

Since Jock was promoted to project management with little training and has difficulty with project budgets and financial reports, he is prone to poor use of resources and lacks budget control. His team also does a lot of rework, since the schedule isn't controlled and milestones are not used to measure progress as work proceeds. Training to understand the project reports and the accounting system is absolutely necessary if projects are to be controlled.

6. People not Held Accountable

While Travers and Jock work long hours in order to meet client deadlines, their staff may arrive late and leave early. It isn't their fault, they just don't know what is expected of them and what to do in most cases. Many workers do not feel part of the project team or committed to their client's objectives because they weren't involved during the project definition phase, and the client's objectives, schedule, and budget haven't been clearly explained to them. They do not "buy in" to a project, since their input was not sought when the project was being defined or planned or when the preliminary scope was developed.

7. Failure to Learn from the Past

In most cases in small firms, project managers fail to hold a project closeout meeting with the project team, to analyze project performance with the client or review actual versus budget performance, productivity, and adherence to schedule. Why does this happen? It should be clear that mistakes from past projects can provide valuable lessons for the future and, if repeated, are serious pitfalls for future projects. Some excuses include not having time because of the need to move on to the next project or not having any budget hours left for such meetings. Sometimes the embarrassment or failure of the project manager to admit to poor performance is the cause. In any case the cost of repeating poor performance can be costly to the firm and demotivating to the entire project team.

HOW DID I GET TO BE A PM ANYWAY?

Most surveyors and engineers don't start their career with project management as their primary career goal. Sometimes we are promoted to the position of project manager because we've done a good technical job, have demonstrated good people skills, and have managed some minor management aspects while working as a member of a project team. Our boss assumes that we've got the skills it takes to be a good project manager, yet we've had little or no project management training. In many small firms, young engineers and surveyors are promoted to project manager positions once they become licensed, in order to take some of the workload from more senior project managers. Most will accept the "battlefield promotion" with some doubt but don't want to deflate their boss's expectations. In addition, they see it as a way to advance their career path and increase their salary, possibly by sharing in bonuses. It almost goes without saying that many people arrive at the position with little or no formal training.

So what do you do having arrived at your new position? Be proactive. Realize that projects are people problems waiting to be solved, and you've been selected to solve them. Project management is also people management, and if you prefer dealing with your computer, solving a complex technical problem, or being far away from the office hustle and bustle in the woods with the survey crew, you should rethink whether or not project management is right for you.

If you decide that project management is the next step in your career, work hard to develop a positive project management attitude. It is a way of leading, thinking, delegating, communicating, and getting things done. It really is application of the engineering method. It is analyzing the problem, gathering data, investigating alternatives, arriving at a viable solution, and implementing it. It also is convincing yourself and others that quality and meeting the client's objectives is the most important thing you can do each day. Finally, it is developing the attitude that "what gets measured gets done."

Why then is this so hard for many engineers and surveyors? I feel that in addition to the pitfalls previously mentioned, there are three primary things, *fear of the unknown*, *a perfectionist attitude*, and the *inability to conceptualize* that lead to problems for project managers.

Since technical education trains people to seek all of the available data and analyze it carefully before proposing a solution, this can lead to fear of the unknown followed by indecision. It is one of the worst things project managers can do. Most project management decisions need to be made relatively quickly and sometimes with less than complete information (data). Our technical training teaches us that people's lives or the public safety often depends on the decisions made by engineers and surveyors. In my experience, I've found that many management decisions can be made with 50–75 percent of the information and, in fact, often full information is not available or there isn't time to analyze all of the data. However, a mistake in a project management decision seldom leads to jeopardizing people's lives or the public safety.

A perfectionist attitude may also be a project manager's Achilles' heel. Many project managers continue to do technical work after being promoted to their new position. As technical problem solvers, they take great pride in their work and often produce creative and unique solutions. This leads them to the attitude that "nobody can do it like I do, so I might as well do it myself." A subset of the perfectionist attitude is the feeling that "it will take too long to train a subordinate to do a certain task (and they won't do it the way I do it), so I may as well do it myself." Both lead to the project manager becoming the "bottleneck" of the project team. The project manager often has too much to do while other members of the team have too little to do. In either case, perfectionists need to remember that clients are not paying for a job to be done to perfection; they are paying for it to be done adequately. This doesn't mean that quality should be sacrificed. To the contrary, the level of quality needs to be discussed with the client and the team, and established by standards and codes, as part of the project definition and planning phases. A quality management protocol needs to be established and followed.

Finally, the inability to conceptualize can be a major issue for new and experienced project managers. Conceptualizing a project means "seeing the forest, not only the trees." Many technical people have the ability to focus on the minutest detail. They generally can see how their solution helps solve a bigger problem, but often they don't see the final solution in terms of the big picture. I believe that this is the reason that many technical people start working on projects before a project is clearly defined or well planned. As stated in Chapter 4, the higher we move up through the management pyramid, the less important technical ability becomes, and the more the ability to conceptualize and see the "big picture" becomes extremely important.

WHY IS PM NECESSARY?

You've done this type of project before, and most of the projects done in small surveying and engineering firms are similar to those that have been done before, so why do you need to manage projects? The project only has a few tasks, and you've done them many times before, so why waste your time with project management, why not just get started and do the work? The reason goes right to the roots of our success—our clients.

First, we need to manage projects in order to clearly communicate with our client that we and they are "singing from the same sheet of music." It doesn't make any difference whether or not we've done this type of survey or site plan or drainage design or bridge hundreds of times or not. Each client is different and each project has different requirements. Project management is a dialogue with the client that opens the lines of communication and demonstrates to the client that we understand their goals and objectives for the project.

Second, project management is necessary in order to communicate with the project team. Many projects, even those done in small firms, require the project manager to coordinate a diverse team. Within the firm, the team often consists of surveyors, engineers, technicians, draftspeople, and administrators. Outside of the firm, we may engage consultants to fill a special project need. They may include wetlands scientists, soil scientists, boring contractors, archeologists, planners, and traffic consultants. Each has a specific role in the project and is a key to the project's success. The role of project management is to coordinate the resources, as needed, for the entire project to be successful. Timing and coordination of consultants work early in the project often is part of the critical path. It can be similar to "herding cats."

The *third* reason to manage projects is to "do it right the first time," to stay on schedule and within budget. The world of consulting surveying and engineering for small firms is very competitive. Surveying, engineering, and permitting are often early project tasks that need to be completed on schedule, since delays can result in late completion of the project. Projects of long duration may need a network schedule in to understand the tasks on the critical path.

Finally, most project managers do not have the luxury of managing only one project. In fact, in our firm, most project managers manage 6–12 projects of varying sizes. This results in a resource juggling act. The main reason that our project managers meet once a week is to determine the allocation of valuable people resources for the following week. This is where project deadlines are discussed and resources are added to critical projects and subtracted from others. Effective regular meetings of project managers are also a key requirement for small firm success. Success is measured by work being completed in a planned and orderly manner and minimizing crisis management when ever possible.

THE PROJECT LIFE CYCLE

Since projects are what small surveying and engineering firms do, it is important to our success that we understand the entire life cycle of a project, from "cradle to grave." Many projects are born from existing clients and require very little upfront or development work. They may arise from a phone call. These are the clients that allow us to invest a minimum amount of nonbillable marketing time up front and proceed to the implementation phase the quickest. Their projects generally involve little nonbillable marketing time and are profitable. Others require a considerable amount of upfront work. (See Figure 5.4).

Pre-Project Marketing Phase (Nonbillable)

Some projects involve a large amount of upfront nonbillable time just to develop the client. It may take years before a real project comes to fruition. Often, before there is a real project, you must invest time to determine if the client and their project(s) are a suitable fit for your firm. Do you have the experience and expertise to do their work? Do they appreciate the high-quality

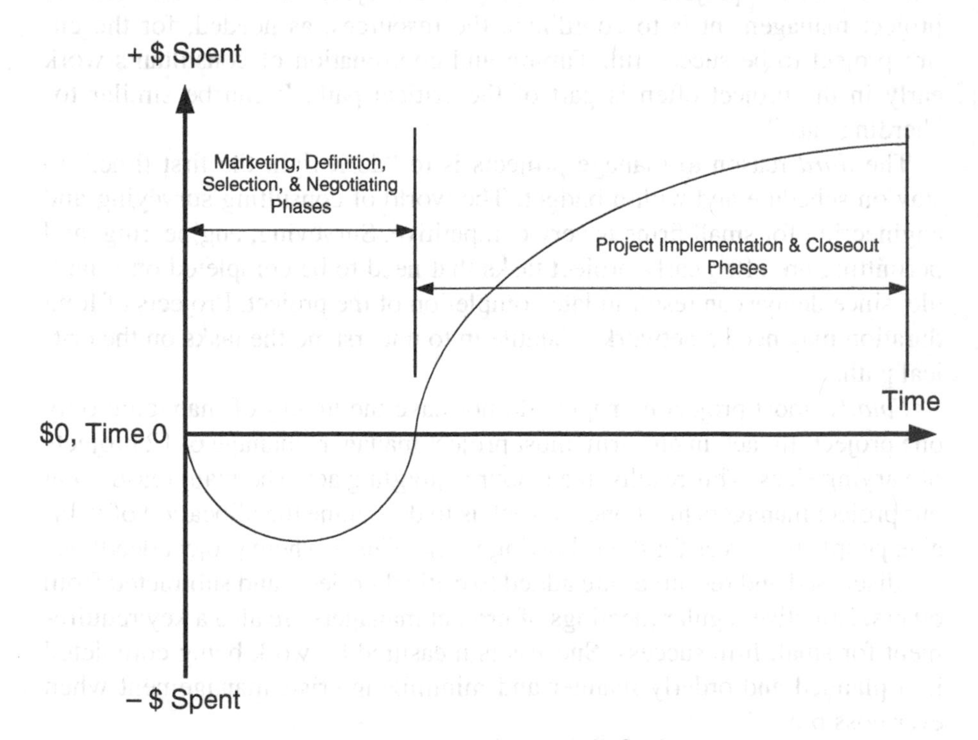

Figure 5.4 The Project Life Cycle.

work done by your firm? Do they pay their bills on time? You may spend a considerable amount of time obtaining the answers to these questions. Nonbillable time spent for client development is generally billed to marketing or overhead, but it needs to be tracked in such a manner that marketing costs to obtain a project can be determined. Firms have been known to spend more in marketing cost than their potential profit for a project. This may be OK for the first project with a new client, but the trend shouldn't continue. Often client development is done years in advance of a project and, since it is a considerable investment, it should always be done for the long term. Notice that I said client development, not project development. More on this in Chapter 9, "Marketing Professional Services."

Project Definition & Scope Phase (Nonbillable)

Once a client determines that they have a problem, a real project arises. Your firm may become involved at this point to help the client further define the problem and develop a preliminary scope, a budget, funding alternatives, and a schedule. Involvement at this phase of a project often gives a firm a competitive advantage, since they helped define the project. It also helps a client who is unfamiliar with the design and construction process understand some of it complexities. In many cases, this upfront work is done by small firms as part of the continuing marketing or client development effort and is "pro bono" nonbillable time. If you can be engaged as a special consultant and get paid for your work during the project definition phase all the better. In my mind, helping the client correctly define the project and establishing an adequate budget for the work is one of the greatest services engineers and surveyors can do.

Consultant Selection Phase (Nonbillable)

Firms that deal with Federal government clients, local municipalities, state agencies, and some private clients are familiar with the consultant selection process. It is a formal process, where the client has defined their problem and seeks the best-qualified firm to do their work. The consultant selection process is often strictly controlled by law or regulations at both the state and federal level. The procedure is known as qualifications-based selection, or QBS. Since a project may only be preliminarily defined, this process seeks qualification statements from firms that have done similar projects and fee is not part of the consideration. Once the most qualified firm is chosen, a fee is negotiated. In the private sector, the process is less formal. Some private clients continue to use the same firm for project after project over a period of many years. Other private clients may "bid" surveying and engineering services, selecting the consultant on a combination of price and qualifications. Efforts in this phase are still nonbillable, since there is no contract yet.

Contract Negotiation and Agreement Phase (Nonbillable)

This is the point where a firm has been selected and it negotiates a contract with the client based on some sort of scope of work. At this stage everything should be on the table. A certain amount of project definition work needs to be done as part of this phase but the extent of the client's problem may not be completely known. The firm estimates man-hours required to do the work, seeks proposals from subconsultants for special tasks, negotiates overhead multipliers or billing rates, and discusses general terms and conditions. If the project is not clearly defined, the scope and agreement may be only for a study that will determine the full extent of the project, with further design and construction phase services to be added later. If a suitable agreement is not reached, the owner may terminate the negotiation and attempt to reach an agreement with their second choice firm. Efforts here are still nonbillable.

Project Implementation Phase (Billable)

Once you've reached this phase, you've moved to the plus side of the ledger and can start to be paid for your services. This is where the "rubber hits the road" in project management. You've spent many hours developing the client and helping define the project. You've gone through the competitive process, been selected, and successfully negotiated an agreement with your client. Now you've got to perform, on time and within budget. The client doesn't want excuses. The first step is to develop the detailed work plan, including the proper accounting procedures, which will allow you to track the project financially, and then hold a project start-up or kick-off meeting with your team. Be sure to include outside consultants if possible.

Once the project has started, the role of the project manger becomes one of management of quality and resources, control, and measuring progress. Hold regular project meetings with the team and make adjustments, such as adding resources if needed, to stay on schedule. The project work plan should specify the level of quality sought for the project and the desired outcomes or deliverables. Project meetings are the time to remind the team that "do it right the first time" not only improves quality but also helps in staying on schedule and within budget.

Project Closeout (Billable)

In my experience, this phase of a project is not done in small firms. This is unfortunate because here is where lessons learned can keep you from making the same mistakes in the future and help improve profits. There are many reasons why project closeout isn't done. Often a project team is broken up upon completion of a project and high priority is given by the project manager to moving on to the next project. Sometimes a project doesn't go well and a project manager doesn't want to be reminded of the details of his failure. Often upper-level management doesn't want project managers spending nonbillable time reviewing a project once it has been closed out and billed. *I don't think there are any good excuses to not close out a project and debrief the entire project team.* Lessons learned can be those things done right as well as those that didn't go so well. In the case of those done right, the project manager has an opportunity to thank the team for a job well done. This can be a great motivator for future projects. Where things didn't go so well, the project manager should approach the closeout meeting with a positive attitude and address situations, tasks, quality, and so forth from the point that this is what we did and how can it be improved the next time? It is important not to attack individuals, since they are likely to become demotivated and see no sense in attending future project closeout meetings.

PLANNING, ORGANIZING, & IMPLEMENTING YOUR PROJECTS

Phase 1–Details of Defining Your Project

How do you define a project? As far as I'm concerned, this is the most important phase of project management. Developing an accurate project definition is key to your project's success. It doesn't matter whether it is a new project for a new client or another project for a long-term client. There are only slight variations if the project is small rather than large, but go through the project definition steps anyway, since the project may turn out to be completely different from your initial impression and the client's initial description of the problem to be solved. The first step is to describe, in as much detail as possible, the nature of the client's problem, as it is known at the time. Is it a survey or an engineering project? Is it a subdivision or site plan? Is it a horizontal project (on or below the ground) or is it vertical (an above-ground structure)? Is it a building or bridge? Is the client seeking services all the way through construction and start-up or does he just need your firm up to the point of obtaining permits? What are the deliverables? How many meetings are you likely to go to? Does the client have the money for the project or does he expect your firm to help in securing grants and other funding? Are outside consultants needed? Who will be on the project team and what are their roles? Be sure to specify the project manager and the number two person in charge, possibly the project engineer or project surveyor. This helps the client understand the hierarchy of the project team. Following is a project definition checklist. It may not be necessary to fill it out for routine projects, but I suggest that if your firm doesn't use such a form, all of the project managers try it out, especially those new to project management. Once it is completed, furnish a copy to each member of the project team, including the client and outside consultants. A project approach was likely developed during the marketing and proposal phase. If it exists, it should be attached to the project definition checklist.

PROJECT DEFINITION CHECKLIST

Project manager: ____________________

Brief description of the client's project and goals: (location and description of the client's problem, funding, and schedule, if known)

____

____

____

____

____

____

____

____

____

Client's name: ____________________

Client contact or project manager: ____________________

Client contact information:

Office phone: ____________________

Office fax: ____________________

Personal cell: ____________________

Email address: ____________________

Physical address: ____________________

____

____

Mailing address: ____________________

____

____

Other stakeholders: (include permit and funding agencies and others who can influence the project outcome)

____
____
____
____
____

ACME project team in addition to PM:

Project engineer: __________________________

Project surveyor: __________________________

Project team members:

____
____
____
____

Consultants: (include boring and test pit contractors, soil and wetlands scientists, archeologists, traffic consultants and contact information, etc.)

____
____
____
____
____

Existing or new client:

_____ Existing Last project: ________________

_____ New

Selection process:

_____ Qualifications-based (QBS)

_____ Price-based proposal

_____ Other: (explain) ______________________

____

Proposal and Scope:

_____ Scope prepared and included in letter agreement

_____ Detailed scope is an attachment to the letter agreement

Detailed scope includes:

_____ Project understanding

(*Continued*)

_____ Assumptions (including owner furnished services, access to site and those furnished by others)

_____ Detailed description of phases and tasks

_____ Description of deliverables

_____ Description of number of meetings expected/included

_____ Description of schedule, including milestones

_____ Description of project manager and project team

_____ Description of fee, hourly/lump sum (include breakdown by phases or milestones and job cost spreadsheet)

_____ Proposal, scope and fee been checked

Agreement:

_____ ACME letter agreement

_____ Other form of agreement (explain) ____________________

____

_____ Signed agreement received

_____ Has the procedure for contract changes been described

Accounting:

_____ Overall budget communicated to accounting

_____ Billing groups established

_____ Invoice timing worked out with client

_____ Project folders/binder started

_____ Project entered into client database

Project work plan:

_____ Has the detailed project work plan been established by the project manager?

_____ Has a project kick-off meeting been held with the entire team?

_____ Have responsibilities of the project team been clearly defined?

Phase 2–Detailed Planning, Strategy and Agreement

Further development of the project definition into a scope of work, which will become part of the agreement with the client, takes place in this phase. Initially, you will develop a preliminary scope based on your understanding and approach to the work. It should be developed in concert with other member of the project team, including outside consultants. The scope format should always contain the following sections:

- *Project understanding*. This is the first paragraph, which describes how you see the project based on what you know to date. It contains some of the information determined during the project definition phase such as the client's understanding of the problem and project. You also may have learned more about the project since the project definition was developed. This is where you can update it. The project understanding also should include a clear statement of the client's goals. An example would be to study and develop alternatives to solve the problems related to a noncompliant waste water treatment plant. It may also contain your understanding of the client's time frame and funding if known. The overall goal of the client should be clearly stated here.
- *Assumptions*. Here you describe the assumption from which will be based your entire project approach. Common assumptions are that the owner will furnish adequate access to the project site. The owner will furnish program requirements or regulatory requirements and copies of all correspondence to date. The site does not contain known hazardous material or that it is being cleaned up by others. In the case of surveying projects, you should state that it is assumed that the bounds for the parcel being surveyed exist or are easily reproduced and that there are no defects in the title. You want to make it clear that you don't plan to survey the entire neighborhood to set a few lot corners or solve a minor deed discrepancy. It also is fair to assume that there are no encroachments, unknown easements, or other significant problems that may be discovered during research. In our part of the country, winter conditions have an impact on the cost of a survey, so if you are assuming that field work will be done in non-winter conditions state it in the assumptions. Where work is being done based on plans that have been done by others, an architect, for example, state that your design will be based on their plans of a specific date.
- *Project approach*. Here is where you actually begin to describe the work that you will do. In our firm, we have templates that detail phases and tasks for typical projects such as bridges or site plans. The traditional phases include:
    - *Pre-design phase* work such as survey and base mapping, geotechnical investigations, wetlands identification, traffic studies, archeological investigation, and any other items that need to be done to provide data for future phases.
    - A *study phase* is often necessary to develop alternate solutions, cost estimates and schedules. It is usually submitted to the owner (and possibly regulators) with a recommended alternate, but the final or preferred alternate should be the owner's choice. If a project is large, is

complex, and has several possible solutions, the scope may not describe future phases, since they may be different and dependent on the alternate solution chosen. For a simple project where the solution is known, you may propose skipping this phase and proceeding directly to the design phases. This may be the case with a simple survey project or site plan.

- The *preliminary design phase* is where the selected alternate or the obvious solution is developed to a level where most of the factors that will affect cost are known. Definitions of preliminary design vary. In some firms or with some clients, preliminary design means developing the design and plans to a level of 50–60 percent complete final drawings. Most of the work done in our firm requires that plans be developed to the level of 75–85 percent complete. At this point, most of the unknown issues have been determined and a detailed cost estimate can be prepared. These plans may also be used to obtain permits from local and state regulatory agencies.
- The *permitting phase* is where your detailed preliminary design is submitted to all agencies that have review and permitting authority over a project. To many clients, this seems like the proverbial "black hole," but if your project team is experienced, the permitting process should be routine. The key to success in the permit process is the coordination of all of the material required for a specific permit and then following through to provide the permitting agency with additional information if necessary or requested. Presenting complete, detailed permit application packages that meet all of the requirements of a permit agency is the best way to obtain permits in the shortest time period.
- The *final design phase* generally follows successful permitting, although some clients only require services through permitting, after which the engineer's services are no longer utilized. During this phase, final changes that may have resulted during permitting are made, additional details may be added to clarify designs and written technical specifications and contract documents are prepared for the "project manual." A final cost estimate is prepared and the project is ready for bidding and construction. I recommend that the final cost estimate contain a 10 percent contingency since the perfect set of plans and specifications has never been prepared and there *will* be changes during construction.
- The *bidding phase* services generally include holding a pre-bid meeting or site visit to familiarize bidders with the project scope, the owner's goals, and the plans and specifications. If the project is a public project and requires public notice and advertising, we prepare the

invitation for bid and instructions for bidders as part of the project manual. Often during the pre-bid site visit questions arise regarding a specific drawing, detail, or material specification. These are clarified by issuing an addendum to all of the bidders so that all are equally informed and still bidding on an equal basis. The next step in the bidding phase is to receive the bids at the advertised time. Again, if the project is public, bids generally are opened and read aloud to those attending the bid opening. At this time, it is important to be sure that all of the bid requirements have been fulfilled, including corporate seal, acknowledgement of addenda, furnishing a bid bond, if required, list of subcontractors and suppliers, and any other information requested in the instructions to bidders or contract documents. Once the bids are received, the "apparent" low bidder is announced, pending analysis of the bids for errors. The final step for the engineer is to carefully check all bids for math errors and make a recommendation to the owner for award of the contract.

- A *construction administration phase service* is the final service provided to many owners. It begins with the owner giving the successful bidder a "notice of award." The form is often prepared by the engineer. Once the contractor is notified, the engineer prepares all of the documents necessary for signature by the owners and contractor in order for the work to start. These are specified in the contract documents contained in the project manual. The first real construction administration service provided by the engineer is the preconstruction meeting. This meeting covers all of the details of how the engineer will be involved with the contractor as the owner's representative. Some of the services include: reviewing shop drawings and submittals, reviewing schedules of values and payment applications, attending regular job progress meetings, issuing clarifications of the contract plans and specifications when the contractor requests additional information (RFIs), visits the site to determine if the contractor is following the design intent, conducting final inspection and preparing the "punch list" of items to be completed prior to final payment, issuing a certificate of substantial completion if necessary, and providing notice to surety of completion and final payment.
- *Project schedule*. The scope of services should also contain a project schedule that is appropriate for the size of the project and the work being performed. A small project, such as a survey or a small site plan, may only require that specific milestone dates be stated and when the work will be started and completed. Larger projects may require a bar chart or Gantt chart schedule, where all of the phases or tasks to be performed are

ID	Topographic Survey Schedule	Start	Finish	Duration
1	Project start-up meeting	7/6/2007	7/6/2007	0d
2	Slack time – waiting for crew available	7/6/2007	7/10/2007	3d
3	Deed research	7/9/2007	7/10/2007	2d
4	Field reconnaisance	7/10/2007	7/10/2007	1d
5	Field survey	7/11/2007	7/19/2007	7d
6	Milestone – Field work complete	7/20/2007	7/20/2007	0d
7	Slack time – crew assigned to new proj	7/20/2007	7/26/2007	5d
8	Field note reduction	7/20/2007	7/20/2007	1d
9	Survey calculations	7/23/2007	7/25/2007	3d
10	Plan drafting	7/26/2007	7/30/2007	3d
11	Plan checking	8/1/2007	8/1/2007	1d
12	Deliver final plan	8/6/2007	8/6/2007	1d
13	Project complete – Closeout & debrief	8/7/2007	8/7/2007	0d

Figure 5.5 Gantt chart schedule.

shown with their respective start and end dates. Gantt charts generally show one phase being completed before another is begun, but if the chart is broken down into detailed tasks, it may show several tasks occurring simultaneously. When this occurs, the task of longest duration is the "critical" task and controls the project completion date. The other tasks have "slack time." This means that these tasks may be completed at any time up to the latest completion date of the critical task. If your project is complex, it may be much easier to use critical path scheduling (CPM) software than to manually try to redo the Gantt chart for each possible scenario (see Figure 5.5).

- *The project team.* The project team also should be described in the scope of services. The project manager, department leaders who will participate in the project, and consultants, as well as other team members should be described. It also may be helpful to include a task and responsibility matrix to graphically show the client which team member has primary and secondary responsibility for each task (see Figure 5.6).
- *Basis of fee and agreement.* The final paragraph of the scope of services should describe to the client how the fee is determined and how invoicing will be done. Generally, there are two ways surveyors and engineers charge for their work. The easiest and less risky method is hourly or time and materials. Here, the owner pays for the actual hours worked on

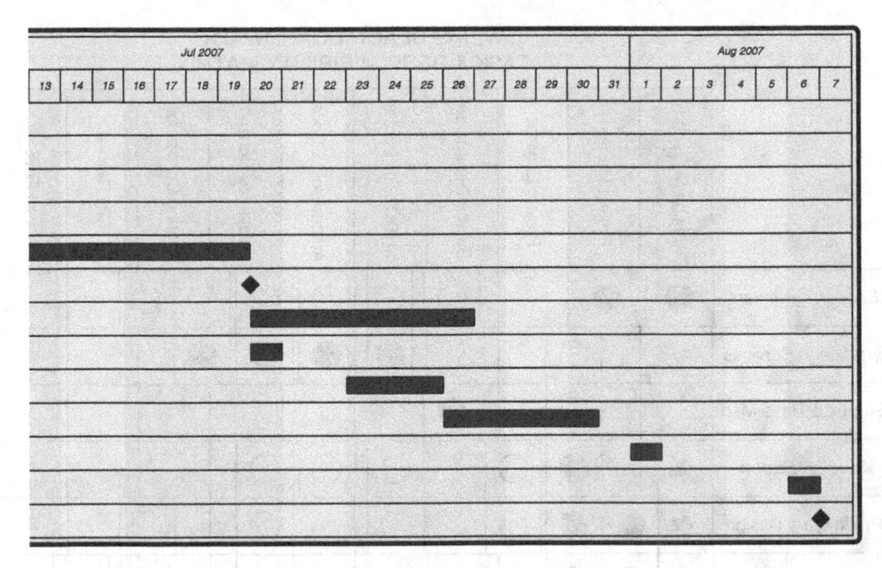

Figure 5.5 (*Continued*)

the project plus reimbursable materials such as mileage, survey bounds, reproduction cost, laboratory fees, rental of special equipment, testing, and consultants. While this method is less risky than the lump sum fee method, described next, the owner generally wants a budget range for the work, and the project manager still needs to notify the owner when tasks are undertaken that are outside of the original scope of services. Be aware of owners who insist that the budget is a "not to exceed" figure, since this requires the surveyor or engineer to take all of the risk. The lump sum fee method provides the greatest risk and also the greatest possible reward. Where the scope of work is known or can be adequately determined the engineer can estimate the number of hours needed for each task in the project and, therefore, determine its fee. By adding up all of the tasks and including the reimbursable expenses, the total fee can be determined. The incentive for lump sum fee work is that, if tasks are completed in less time than anticipated, greater profit is realized. Lumps sum fee work also presents the possibility for greater loss if the number of hours for a task exceeds the amount planned, the firm takes the loss. Many owners prefer lump sum fee proposals, since it sets the fee for their project and helps them establish their project budget. In order to minimize risk, a lump sum fee should only be proposed when the scope of work is well known, hours easily tracked, and the firm has experience with the type of project. There are variations to the preceding fee determination methods, including the hourly cost times a multiplier

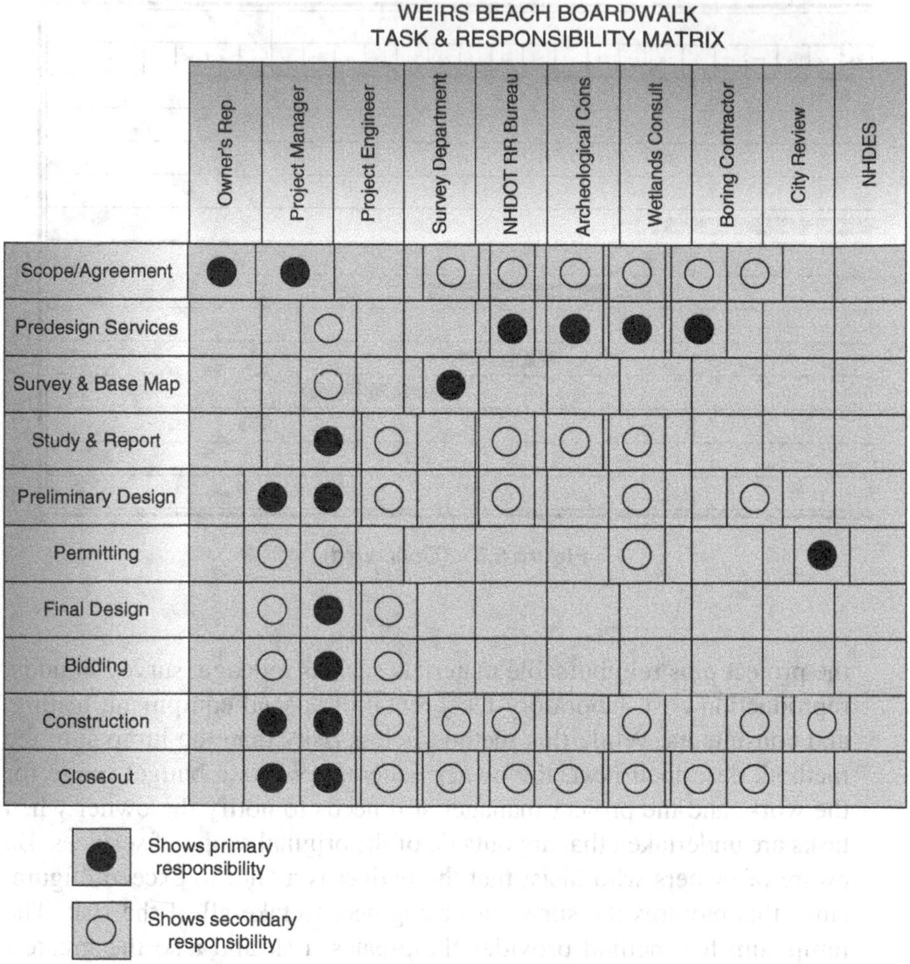

WEIRS BEACH BOARDWALK
TASK & RESPONSIBILITY MATRIX

	Owner's Rep	Project Manager	Project Engineer	Survey Department	NHDOT RR Bureau	Archeological Cons	Wetlands Consult	Boring Contractor	City Review	NHDES
Scope/Agreement	●	●		○	○	○	○	○	○	
Predesign Services		○			●	●	●	●		
Survey & Base Map		○		●						
Study & Report		●	○		○	○	○			
Preliminary Design	●	●	○		○		○		○	
Permitting	○	○					○			●
Final Design	○	●	○							
Bidding	○	●	○							
Construction	●	●	○	○	○		○		○	○
Closeout	●	●	○	○	○	○	○	○	○	

● Shows primary responsibility

○ Shows secondary responsibility

Figure 5.6 Task and responsibility matrix.

method used by some state departments of transportation. In this type of fee arrangement, the multiplier and profit may be limited to certain allowables. In our firm, we use a "job cost" spreadsheet to determine the fee for every project. This gives us a standard format for everyone's use and allows the opportunity of varying the multiplier and profit if needed. See the sample in Table 5.1.

Often only part of the overall project approach is authorized by the owner at the beginning of a project because there are too many unknowns. Many engineering projects begin with the owner only authorizing the study and

Table 5.1 PROPOSED PROJECT FEES

SAVE UNDER MARKETING OR JOB # BEFORE ENTERING VALUES

	Change this column only
Project Title:	**Pleasant Way Road Reconstruction**
Client:	**Town of Pleasantville**
Preparer:	**John Q. Smart, P.E. - ACME Engineering & Surveying**
Date:	**January 01, 20XX**

Phase ID	Billing Group Code	Project Phase	Fee	Man-Hours	Comments
09	001	Pre-Design Survey & Geotech	$	0	
10	002	Schematic Design	$	0	
11	003	Preliminary Design	$	0	
15	004	Permitting	$	0	
12	005	Final Design	$	0	
13	006	Bidding	$	0	
14	007	Construction Observation	$	0	
	TOTAL		$	0	

report phase, with future phases depending on the selected alternative and/or funding for the project.

Remember that submitting your preliminary scope to the owner is just the beginning of the negotiation of your agreement, and it may require several iterations before both of you agree upon the work to be done and the fee for services. I often leave the fee out of the first submission of the scope in order to be sure that the owner is in agreement with the work to be done without distracting him with the fee. Once agreement on the scope is reached, the fee is easily calculated and added to the final document.

Upon completion of the negotiation of the scope of work and fee, the final step is the actual agreement document itself. It contains a description of the work, and the detailed scope of services is attached. It also contains the standard terms and conditions that govern the contract between the firm and the owner. These should be industry standard terms and conditions modified to your specific situation but they should remain constant. Occasionally, an owner will reject your terms and conditions and insist that theirs be used. Beware of owner-generated terms and conditions. Often they have been developed to cover purchasing products not professional services. Terms such as "guarantees and warrantees" are not appropriate for a professional service agreement. When in doubt about owner-generated terms and conditions seek legal advice or have them reviewed by your professional liability insurer. This generally is a free service that your insurance company provides, and we have used it many times. The goal for you and the owner should be to have terms and conditions that are fair to both sides and insurable.

Phase 3–Implementation & Execution

The implementation phase of a project is where project managers have the first opportunity to show their stuff. Unfortunately, it also is where many project managers set up their projects for potential failure. In my experience, the main reason is that project managers just don't take the time to develop the project definition into a detailed and effective scope of services, followed by a work plan. Being technical people and in most cases having risen from technical roots as a project engineer or surveyor, project managers are just too anxious to roll up their sleeves and get involved in the technical solution of the client's problem. If a project is to be successful, it must have a well-conceived scope, followed by a work plan at project implementation, and it must be conveyed to all members of the project team.

The Detailed Work Plan In most cases, the work plan is too detailed for the client, but the essence has been stated in the project scope of services. The goal of the work plan is to develop an overall approach to implementing the project and accomplishing the client's goals. Is it a small project with a team of two or three people that can be completed in a few days, or is it a large project which may take months or even years to complete? In the first case, a project manager may be able to handle many small projects and keep the work flow for all of them going concurrently. This can be similar to "herding cats," and it takes an experienced, well organized, project manager who delegates well to team members and sees his/her primary job as client liaison, quality management, implementation, control, follow-up, and follow through. In the case of a large project with a large team, several consultants, and a long duration, actual management of the work may be easier than managing many small projects, but a significant commitment to planning implementation as well as managing the project is still required. A difficultly arises when the manager of a large project also manages several small projects and many of his resources are shared between the big and small projects. Priority conflicts often arise. Management of multiple projects is no small task, and an entire section will be devoted to it later in this chapter.

The next step in implementing a project is for the project manager to draft a work plan and circulate it among the project team for review, input, and comment. Input from all team members is important. Be sure to include consultants and the accounting department. The work plan may be the first opportunity for team members to see a detailed description of the project and their role. It also provides an opportunity to communicate their input and suggestions to the project manager of possible glitches that they see in the project scope or suggestions for improving efficiency, quality, and boosting team productivity and motivation. Project team members also need to communicate (and coordinate) their availability for commitment to the project early on in the work plan process. A key member of the team who isn't available until other commitments are fulfilled could impact meeting milestones, deliverables and may impact the overall success of the project. The work plan must contain a description of all of the tasks and phases (including accounting codes) and who on the team will perform them. The work plan should be a detailed description of how the work will actually be done, including time allocations and schedules for each task to be completed. Here is where a schedule, which is more detailed than the one presented in the scope of services, is developed. Often, before significant work can begin, pre-project work by outside consultants or contractors is necessary. Time for this work must be programmed into the project schedule, and assignments for other members of the project team must be made so that they are available when needed. A group of tasks, when completed, generally constitutes completion of a certain phase of the project and often is a milestone, where input from others is necessary. At this point there may be deliverables, review meetings, invoicing, selection of alternatives, and possibly adjustment of scope, schedule, and budget. The project also may "go on the shelf" for a period of time, while the client, regulators, funding agencies, or other stakeholders conduct their review and provide input. The project manager must be aware that work which takes place on the project during this time may (possibly to catch up and get back on schedule) be risky and could require rework. It may be best to assign the project team to another project while this is happening. Provisions for adequate review and quality control must be contained in the work plan as well as provisions for adjusting the project schedule if the reviews do not take place in a timely manor.

As previously mentioned, the work plan should also contain a detailed project schedule, which in most cases can be a list of start and completion dates for the tasks in each phase. A bar graph or Gantt chart schedule showing the duration of each task and milestones is usually the result. Complicated projects or those of long duration may require critical path scheduling. This is normally done using a project-scheduling program like

MS Project to identify those tasks, which make up the critical path. The important thing for a project manager to remember is that a critical path schedule is no good if it is not updated regularly. Regular updating of a critical path schedule can become a very time-consuming part of project management.

Project accounting and involvement of the accounting department should also be included in the work plan. Invoicing milestones must be known and any requirements of the client for a particular invoice format or particular invoice procedures and submission dates must be known if project cash flow is to be optimal.

Once input has been received on the work plan, it should be revised and a project start-up or kick-off meeting should be held with the entire project team if possible. The meeting should review all of the aspects of doing the work and input from the team may result in another revision before the plan is issued as a final document.

Finally, the project work plan should clearly describe the deliverables and the overall process of ensuring project quality. Are the deliverable(s) a survey, subdivision plan, report, permits, or drawings and specifications? The work plan should describe how and who checks calculations, estimates, reports, drawings, and specifications. It also should describe who reviews important correspondence before it leaves the office and how records of transmittals are kept. If drawings or reports are to be sealed, the responsible person should be named. If the project drawings involve multiple disciplines, there may be a surveyor and several engineers who seal individual drawings in the drawing set. In our firm, we have initiated a quality control measure whereby a senior-level person, who has not been involved in the project, reviews the completed drawings and specifications at significant phases or milestones and before the final release of the documents.

It is important to be sure that all work is completed in sufficient time before submittal deadlines to allow for adequate final checking and revisions. If the quality control system works, there should only be minor issues to deal with as deadlines or milestones approach.

Remember the old project manager's saying, *"Plan the work, then work the plan."*

If the project work plan was carefully developed and implemented, executing the project becomes a planned routine function for the project manager and members of the team. No work plan is perfect and adjustments will be required along the way, but having a work plan means implementing the project in a planned mode rather than in a crisis or reactionary mode.

[Project Name] Work Plan Checklist

Project Manager: [name]
Owner: [name]

Risk & Problem Management

- ❑ The project team has invested time and energy into identifying all project risks.
- ❑ Contingency funds or time have been allocated by management for problems.
- ❑ A problem log has been developed to manage known problems and is accessible to all project team members.
- ❑ Every risk and problem in the log has someone responsible for managing it.
- ❑ There is a plan in place for continuously identifying and responding to new problems.

Work Plan Breakdown

- ❑ Tasks and responsibility have been identified to produce every deliverable in the scope of work.
- ❑ The work breakdown structure (phases and tasks) for the project is consistent with company standards and guidelines and/or templates for similar projects.
- ❑ The project team participated in building the scope or has reviewed and approved it.
- ❑ Every phase and task in the scope has a strong, descriptive name that adequately describes the work.
- ❑ Every phase and task in the scope has a beginning, an end, and clear completion criteria.
- ❑ Project management tasks are included in the work plan.
- ❑ Phases and tasks have been broken down to a level that enables clear responsibility to be assigned.
- ❑ The work plan has been reviewed by the project team.
- ❑ The structure of the work plan has been independently evaluated to ensure the phases and tasks are meaningful to all stakeholders.

Phase & Task Sequence

- ❑ All work phases and tasks have predecessor phases or tasks identified and relationships are Gantt chart.

(*Continued*)

- ❑ The team has reviewed all predecessor-successor relationships to ensure there is clear understanding, none missing and that none of the relationships are unnecessary.
- ❑ All deliverables have predecessor-successor relationships defined.
- ❑ External schedule constraints are represented by milestones.

Fee estimating

- ❑ Wherever possible, historical data has been used as the basis for estimating fee.
- ❑ Ball park estimates have only been used for initial screening and are not the basis for any phases, tasks, or consultant fees.
- ❑ Estimates have been prepared by a PM or principal who understands how to perform the work and who understand the constraints of the people who will perform the work.
- ❑ Buy in by all team members has been received.
- ❑ Job cost estimates include all estimated labor, consultants, reimbursable, and administrative costs.

Scheduling

- ❑ A critical path analysis has been performed to identify critical path tasks and schedule float.
- ❑ Resource leveling has been applied to ensure that the schedule represents a realistic allocation of personnel and other resources.
- ❑ The schedule is based on realistic assumptions about the availability of project personnel.
- ❑ Portions of the schedule that contain many concurrent tasks have been evaluated for risk.
- ❑ The cost-schedule-quality equilibrium is realistic and acceptable to the project manager and owner.

Consultants & subcontractors

- ❑ Consultants and subcontractors have signed contracts with specific scopes of work.
- ❑ The work to be performed by consultants and subcontractors is integrated into the scope and schedule.
- ❑ There are specific milestones and activities planned for monitoring consultants and subcontractors.

Approvals

- ❑ The detailed action plan has been presented and approved.

Phase 4–Controlling the Project

Managing Scope Changes This phase of the project manager's work is generally the least liked and the most difficult to get a project manager to do. It also is the most important when it comes to delivering a high-quality work product on time and within budget. One of the most difficult, but important, things for a project manager to do is to be intimately familiar with the project scope and constantly aware of requests for new tasks that were not included in the original scope and, therefore, subject to adjustment of the scope and additional fees. This is known as *"scope creep."* The project manager can receive a lot of help in this area by training the team to be on the look out for scope creep. What often appears to be a few simple requests by a client to study another alternative, revise a cost estimate, attend a few more meetings, prepare easement descriptions, or include a survey layout when none was planned, can accumulate into a large amount of time and money that wasn't planned and may have serious impact on the project budget. When scope changes arise, many project managers contact the accounting department to see "How we are doing on the budget?" There is a tendency with some project managers to ignore minor scope changes if they can be absorbed within the original project budget, that is, " a few hours more than I originally planned will have little impact on project profitability" or " I can make it up on future phases." Other project managers *intend* to "talk to the client" about scope creep but do the additional, out-of-scope work first, which puts them in a very poor position to be paid after the work is completed. A typical client's response is "I thought that the task was included in the original scope and if I had known it was outside of the scope I wouldn't have requested that the task be done." In our firm, all project managers are required to communicate scope changes to their clients immediately and *before* the work is done. Included in the communication is a fee estimate and change of schedule for the additional work. This is memorialized in a contract amendment known as an "Additional Work Authorization," or AWA.

Monitoring the Budget Monitoring the project budget is an important task for a project manager. Most firms have project accounting software where project budget hours and fee can be entered at the beginning of the work. This task should be undertaken at the time of the project kick-off meeting and is the reason that the accounting department should be involved in the meeting. Our software allows those who work on a project to enter their time billed to the project as they work on it. Once a week, after all of the time report entries are completed, a work-in-process report (WIP) report is printed and distributed to the project managers. This report is an overall status of all of a project manager's work and shows the total fee expected, the amount

ACME Engineering & Surveying

ADDITIONAL WORK AUTHORIZATION #

Date: **Project No:**

To:

From: **PM:**

Re:

In accordance with our original Letter Agreement dated ____________, you hereby authorize us to proceed with additional services as follows:

Deliverables:
Meetings:

Compensation shall be on an hourly basis under the fee schedule in effect at the time services are performed; an estimated fee is __________, plus reimbursable expenses.

The work shall be performed and completed within __________ weeks from notice to proceed.

Upon the signature of this authorization, this service and the General Provisions shall become part of the original agreement identified above. If you agree with these arrangements, we would appreciate your returning a copy of this agreement signed by an authorized representative.

ACME Engineers & Surveyors, Inc.

By: ______________________________________

John Q. Smart, P.E.

Title: President

<Client Name>:

Accepted this _______ day of ________________________ 200×

By: ______________________________________

Title: ______________________________________

GENERAL PROVISIONS

(Terms and Conditions)

Access to Site

Unless otherwise stated, the Client will provide access to the site for activities necessary for the performance of the services. ACME Engineers & Surveyors, Inc. (ACME) will take precautions to minimize damage due to these activities, but has not included in the fee the cost of restoration of any resulting damage.

Fee

The total fee, except when stated as a fixed fee, shall be understood to be an estimate, based upon Scope of Services. Where the fee arrangement is to be on an hourly basis, the rates shall be in accordance with our latest fee schedule. Reimbursable expenses will be billed to the Client at actual cost plus 15 percent.

Billings/Payments

Invoices for services will be submitted monthly and are due when rendered and shall be considered PAST DUE if not paid within 30 days of the invoice date. A monthly service charge of 1.5 percent of the unpaid balance (18 percent true annual rate) will be added to PAST DUE accounts. If the Client fails to make payment when due and ACME incurs costs to collect overdue sums, the Client agrees that all such collection costs shall be payable to ACME. Collections costs shall include, without limitation, legal fees, collection fees and expenses, court costs, and reasonable ACME staff costs at standard billing rates for ACME's time spent in collection efforts. If the Client fails to make payment when due or is in breach of this Agreement, ACME may suspend performance of services upon ten (10) calendar days' notice to the Client. ACME shall have no liability whatsoever to the Client for any costs or damages as a result of suspension caused by any breach of this Agreement by the Client. Upon payment in full by the Client, ACME shall resume services and the time schedule and compensation shall be equitably adjusted to compensate for the period of suspension plus any other reasonable time and expense for ACME to resume performance. Retainers shall be credited on the final invoice. If the Client fails to make payment to ACME in accordance with the payment terms herein, this shall constitute a material breach of this Agreement and shall be cause for termination of this Agreement by ACME.

Indemnifications

The Client agrees, to the fullest extent permitted by law, to indemnify and hold ACME harmless from any damage, liability or cost (including reasonable attorney's fees and costs of defense) to the extent caused by the Client's negligent acts, errors or omissions and those of his or her contractors, subcontractors

(*Continued*)

or consultants, or anyone for whom the Client is legally liable, and arising from the project that is the subject of this agreement.

Risk Allocation
In recognition of the relative risks, rewards, and benefits of the project to both the Client and ACME, the risks have been allocated so that the Client agrees that, to the fullest extent permitted by law, to limit the liability of ACME to the Client, for any and all claims, losses, expenses, damages of any nature whatsoever, or claims expenses from any cause or causes, including attorneys' fees and costs and expert witness fees and costs, so that the total aggregate liability of ACME to the Client shall not exceed $50,000, or ACME's total fee, whichever is greater. It is intended that this limitation apply to any and all liability or cause of action however alleged or arising, unless otherwise prohibited by law.

Termination of Services
Either party may terminate this agreement for cause upon giving to the other party not less than seven (7) calendar days' written notice for: substantial failure by the other party, assignment of this agreement or transfer of the project to any other entity without prior written consent, suspension of the project by the Client for more than ninety (90) days, or material changes in condition that necessitate such changes. In the event of termination, the Client shall pay ACME within 15 days for all services rendered to date of termination, all reimbursable expenses, and reimbursable termination expenses.

Ownership of Services
All reports, drawings, specifications, computer files, field data, notes, other documents, and instruments prepared by ACME as instruments of service shall remain property of ACME. ACME shall retain all common law, statutory and other reserved rights, including the copyright thereto.

Applicable Law
Unless otherwise specified, this agreement shall be governed by the laws of the State of New Hampshire.

Claims and Disputes
In an effort to resolve conflicts that arise during design or construction of the Project or following completion, the Client and ACME agree that all disputes in excess of $5000, arising out of or relating to this agreement or the Project shall be submitted to nonbinding mediation unless the parties mutually agree otherwise. Disputes under $5000 shall be decided by Small Claims Court.

Pollution Exclusion
Both parties acknowledge that ACME's scope of services does not include any services related to the presence of any hazardous or toxic materials. In the

event ACME or any other party encounters any hazardous or toxic materials, or should it become known to ACME that such materials may be present on or about the jobsite that may affect the performance of ACME's services, ACME may, at its option and without liability for consequences or any other damages, suspend performance of the services under this agreement until the Client has abated the materials and the jobsite is in full compliance with all applicable laws and regulations.

Additional Services
Additional services are those services not specifically included in the scope of services stated in the agreement. ACME will notify the Client of any significant change in scope that will be considered additional services. The Client agrees to pay ACME for any additional services in accordance with our latest fee schedule.

Design without Construction Phase Services
It is understood and agreed that ACME's services under this agreement do not include project observation or review or any other construction phase services, and that such services will be provided for by the Client. The Client assumes all responsibility for interpretation of the Contract Documents and for construction observation. The Client waives any claims against ACME that may be in any way connected thereto. In addition, the Client agrees, to the fullest extent permitted by law, to indemnify and hold harmless ACME, its officers, directors, employees, and subconsultants (collectively, ACME) against all damages, liabilities, or costs, including reasonable attorneys' fees and defense costs, arising out of or in any way connected with the performance of such services. If the Client requests that ACME provide any specific construction phase services, and if ACME agrees in writing to provide such services, then they shall be compensated for as Additional Services as provided above, and such services shall become part of this agreement.

billed to date and the unbilled work-in-process. Work done by outside consultants and reimbursable expenses also can be shown. While the WIP report is a good quick overall view of a project manager's project status, it doesn't give detail for each project. The accounting system also should be able to produce a project budget reports for each project which breaks down a project into budgeted man-hours and fee for each phase or task and shows the actual man-hours and fee expended to date as a comparison. This is a key report for a project manager to determine how each project is doing. If the project budget report doesn't provide enough detail regarding who did what and when, the system should be able to produce the next level of detail, which shows who actually is charging hours to the project and the associated tasks. This report is helpful when someone suspects that time is being improperly billed to a project or billed to the wrong phase or task. Monitoring the invoicing cycle and collections is also the project manager's responsibility.

Monitoring Productivity and Schedule Monitoring productivity and the project schedule is the third project control task. This can be particularly difficult for a project manager who is managing many small projects. Things often get out of control with small projects before the accounting reports show that it has happened. Then it's too late. A key to managing many small projects is balancing your resources and having a *"do right the first time"* attitude. This requires the project manager to have a good understanding of the personal productivity of each person on his team as well as their commitment to other projects which may have higher priority. It also requires close communication and understanding of scope and priority by the project team. A staff utilization report, which shows staff members' percent billable time and where nonbillable time is being spent, is useful for the project manager who is trying to get the most productivity out of his team. Overall, management of many small projects is a very intuitive task, which requires an experienced project manager.

Monitoring the project schedule requires the project manager to be aware of each of the project's milestones, deadlines, and deliverables and the performance of the team toward meeting them. Schedules for small projects are generally bar charts or Gantt charts. A critical path schedule may be required for a large project. The schedule generally changes as a project progresses. If the project manager is to keep the project under control, he/she must be aware of these changes and update the project schedule so that the entire team is aware of the changes and how they affect their work. A detailed discussion and examples for managing multiple projects can be found in *Project Management for Dummies* by Stanley Portny.

Managing Quality Each firm should have standards that are used firmwide to ensure that the work product, whether a letter, report, cost estimate, drawings, or specifications, meets a high level of quality. Keep in mind that the client is owed the level of quality he contracted for and nothing more, but certainly nothing less. Quality standards, applicable codes, permit agency requirements, and the level of quality and detail should be carefully detailed in the project work plan and discussed at the project kick-off meeting. The level of quality also should be discussed as the project progresses and shouldn't be relegated to final "red line" mark-ups by the project manager. The goal of each team member should be to produce the highest-level work product they are capable of and to minimize or even eliminate "red line" mark-ups. If team members are experienced, trained, and mentored to understand the level of quality required by the firm's standards and any special requirements of the project that they should produce a work product that requires very little review and rework. It also is helpful to have a person who is not involved with a particular project, review the preliminary work product in the early stages to ensure that the design concept or alternates being studied meet the owner's and other requirements. Constructability also should be reviewed early in a project by the firm's construction staff or in conjunction with equipment suppliers or local contractors who are familiar with the type of construction proposed.

Phase 5–Closeout

The most important phase in a project is closeout. Here is where the final deliverables have been completed, the last inspection made, the final meeting held, the last invoice sent, and the team moved on to the next project. This is where most firms stop, but there is more to be done if we are to learn from our experience. Excuses for not doing project closeout include; not enough time, too busy, and the team has already moved on to the next project. Maybe the budget already has been used up and doing closeout functions and holding a project closeout meeting only adds more nonbillable time to the project account. In essence, reviewing an unsuccessful project may be painful for all who were involved.

Once the project manager receives final feedback from the client and the project accounting is complete, he should hold a debriefing meeting with the entire project team. The meeting should discuss each aspect of the project from beginning to end with emphasis on what went right or wrong and what can be done better the next time. The project manager should share the final accounting so that the team knows if they contributed to making a profit or sharing some responsibility for a loss. He also should discuss whether the schedule was met and if scope creep was dealt with effectively. Some questions that may be asked are:

- Was the project started on the right foot? Was a kick-off meeting held?
- Was the team kept informed of changes in the scope, schedule, and budget?
- Were regular project meetings held?
- Where regular job progress reports sent to the client?
- Did the project manager communicate regularly with the individual team members? Did motivation remain high throughout the project?
- Was the team well informed about the project's goals and objectives?
- Were the team resources used effectively? Did everyone on the team know "who's on first?"
- Did each phase of the project flow smoothly?

- Where did rework happen? Was it minor or significant? What caused it, and how can it be avoided in the future?
- How was cash flow? Did the client pay the retainer? Were invoices paid promptly? Did the PM resolve contested invoices effectively?
- Did the project manger communicate effectively with the client? Were scope creep and its impact on schedule and budget discussed before the work was done?
- Did the PM deal with client issues immediately and to the client's satisfaction?
- Were regular progress meetings held with the client and the team?
- What went well with the project and should be done again? What went wrong?
- Do we want to work for this client again?

Some input for the project closeout meeting can be obtained from a client questionnaire, if one is sent. Within our firm there are mixed feelings about the client questionnaire. Should it be sent out for all projects and to all clients? Should it be sent with invoices? Should it be sent periodically throughout the project or just at the end? These issues aside, there may be something to be learned by receiving written feedback from a client. Maybe the client was not the type of person to deal with problems in person but is comfortable writing them in a questionnaire response. It may be helpful if the questionnaire is returned to someone other than the project manager. The client may have had a problem with the project manager that he'd like to discuss with someone else.

Following the project closeout meeting and team debriefing, a "lessons learned" memo should be written by the PM, distributed to each member of the team, and placed in the file. Client follow-up may also be necessary.

[Project Name] Closeout Report

Project Manager: [Name]
Owner: [Name]

Project Goal

Project Objectives & Results

	Objectives	Results
1		
2		

Scope Comparison

Additional work added (list AWAs)

Decreased scope

Cost Performance

Cost Categories	**Approved**	**Actual**
Internal Labor hours		
External costs		
Labor (consultants, contract labor)		
Equipment, lab fees, permit fees		
List other costs such as travel and reimbursables		
Explanation of cost variance		

Schedule Performance

	Approved	**Actual**
Project completion date		
Explanation of schedule variance		

Major Obstacles Encountered

1
2
3
4

Lessons Learned That are Relevant to Future Projects

1
2
3
4
5

So you've taken the step into project management, and earlier I explored some of the issues that can lead to project management failure. I'm sure many people can identify with the examples of Travers Topgun and Jock Young.

In summary, what do you need to do to be an effective project manager? You need to effectively manage people, processes, and things. You need to manage people, including clients, the project team, and stakeholders that may be outside of your control. You need to manage processes, including defining project definition and plan and how it will be done. You need to set the level of quality to that which meets the client's objectives and satisfies the requirements of all professional standards and codes. You need to control the project by managing the schedule and budget, and finally you need to ensure that the project's objectives are met in terms of deliverables or other expected outcomes.

The project management system previously described has evolved from almost 40 years of experience in small surveying and engineering firms. It is not a system of financial software or critical path schedules. It is an organic system that encompasses the entire project, as previously mentioned, from cradle to grave. There are many other systems and entire books have been written on project management. See the reference section for this chapter for details on where to find them. There may be another system that works better for you or your firm. The important thing is not which system is best, but that project management is embraced by upper-level management, project managers, and the project team and done consistently in your firm.

6

ETHICS & PROFESSIONALS

"The engineer shall apply his specialized knowledge and skill at all times in the public interest, with honesty, integrity and honor."

—William Wisely

"The bottom line in engineering ethics is the idea that engineering ought to be aimed at the good of humanity, and that individual engineers ought to be using their skills to improve the lot of humanity."

—Deborah G. Johnson

INTRODUCTION

It sometimes is said that ethics is what you do when only God is looking. In today's world of athletes claiming they don't participate in doping, politicians being lavishly entertained by lobbyists with no apparent shame, and students cheating on exams being almost an accepted practice, are ethics applicable to the engineering and surveying professions anymore? How does your employment by a company, governmental agency, or corporation affect your independence and ethical practice? Can we expect companies to have ethics or behave ethically in all situations? When we have an ethical dilemma do we ask what is legal before even considering what is ethical? Some older practicing engineers and surveyor may recall the days when codes of ethics forbid professional advertising, required that we treat fellow professionals with respect, and forbid bidding against each other for projects. The U.S.

Supreme Court struck down these provisions in most professional codes of ethics in the 1970s, and the practice of most professions has since changed considerably.

Before we examine ethics in depth we need to understand core values, morals, and the basis for moral behavior. We'll explore what defines a profession and why ethics appears to be more applicable to professionals than nonprofessionals. We'll also examine ethics and business issues related to things such as whistle blowing, conflicts of interest and the effect of corporations on an individual's ethical behavior. We will discuss whether corporations can have ethics and behave ethically.

If we assume that moral and ethical behavior is still required by professionals and codes of ethics are still applicable to professional practice, what are our ethical obligations to our clients, employers, fellow professionals, and the public in general?

Codes of ethics for engineers and surveyors will only be referred to in general in this chapter. For the National Society of Professional Engineers code see www.nspe.org, and for the National Society of Professional Surveyors creed and cannons see www.nspsmo.org. In addition, specific cases of ethical review and two videos on ethics, *Gilbane Gold* and *Incident at Morales*, are available through the Murdough Center for Engineering Professionalism at the Texas Tech College of Engineering and the National Institute for Ethics in Engineering at www.niee.org.

AN ETHICAL DILEMMA

Roger Boisjoly, an engineer at Morton Thiokol, the NASA contractor for the solid rocket motor on space shuttle Challenger, faced ongoing resistance to his concern with O-ring seals and hot gas blow-by when the temperature was cold. As we all know, failure of the O-rings led to the Challenger disaster, which occurred January 28, 1986. Following the disaster Boisjoly testified before the Presidential Commission, which investigated the accident, and subsequently took his ethical "lessons learned" onto the lecture circuit. The following discussion is summarized from his 1987 lecture at MIT.

In January 1985, investigations of O-ring seals began as the result of Roger Boisjoly's postflight inspections of evidence that combustion gas blow-by compromised the two O-ring seals of an earlier space shuttle flight. Morton Thiokol, NASA's contractor for the solid rocket motor, conducted the investigation under Boisjoly's leadership that concluded that cold prelaunch temperatures caused shrinkage of the O-rings and that the gap created would lift off of the surface, which was to be sealed.

Following an April 1985, flight it was found that the primary O-ring seal had experienced blow-by in three places and that the secondary seal also had eroded because of hot gas blow-by. Boisjoly was now very concerned and presented the results of tests, which showed that the O-rings did not lose contact with the surface to be sealed at 100 degrees Fahrenheit, lost contact for 2.4 seconds at 75 degrees, and lost contact for more than 10 minutes at 50 degrees. Based on the test results and because of the cold prelaunch temperature of the April 1985 flight, Boisjoly postulated that the primary O-ring never sealed during the entire 2-minute flight. He mentions that everyone in the program now knew the effect of low temperature on the joint sealed by the O-rings and that Morton Thiokol should be very concerned with the problem.

Boisjoly attempted to form a team to further study the seal erosion problem, but it failed, he claimed, because of a lack of support from management. As a result he wrote a memo on July 31, 1985 to the vice president of engineering at MT stating, "It is my honest and very real fear that if we do not take immediate action to dedicate a team to solve the problem, with the field joint having the number one priority, then we stand in jeopardy of losing a flight along with the launch pad and facilities." Subsequently, a Seal Task Team was formed, but it was frustrated because of the lack of management support.

NASA's Marshall Space Flight Center, also being aware of the O-ring issue, asked Morton Thiokol to send Boisjoly to the SAE Conference in October 1985 to solicit help from experts regarding the O-ring seal. Boisjoly continued to be frustrated by the lack of management support for his team and wrote several memos to MT management regarding the problem but never received a response. This led to the events of January 27, 1986, the evening before the Challenger launch, where several meetings were held to discuss whether or not to postpone the launch since temperatures the following morning were forecast to be 18 degrees. Boisjoly recommended against the launching if the temperature was below 53 degrees, but MT management, who recommended the launch anyway, overruled him. NASA was anxious to receive MT's recommendation, and the launch proceeded with the ensuing disaster. On the evening of January 27th Boisjoly wrote in his notebook:

> I sincerely hope that this launch does not result in a catastrophe. I personally do not agree with some of the statements made in Joe Kilminster's (MT's vice president) written summary stating that SRM-25 is ok to fly.

Following the disaster and his testimony before the Presidential Commission investigating the accident between February and May 1986, Boisjoly received his old job back but had essentially no responsibility. He suffered psychologically, was medicated, and took an extended sick leave. Essentially, his career as an engineer at Morton Thiokol was over.

His final message to the students at MIT was that they should be ready to apply ethics and engineering knowledge and to have the conviction and moral responsibility to stand up and expose questionable practices in industry, which may lead to unsafe products. In essence, Boisjoly was saying, stand up for the primary ethical cannon of the engineering profession, "the engineer's first responsibility is to the public safety, health, and welfare."

Did Boisjoly do all within his power as an employee of MT to affect the launch decision? Were his actions ethical? Did he go far enough by just objecting to the decision to launch? Were business ethics involved here, and if so what role did they play? This chapter explores all of these questions in an attempt to bring our ethical dilemmas and responsibility as professional into perspective.

EXAMPLES OF UNETHICAL BEHAVIOR

Prior to beginning an in-depth study of ethics, let's explore some well-known examples of unethical behavior that also present more ethical dilemmas.

- *Incorrectly billing time*. Yes, it is done and clients are being cheated. I know a college professor who left practice in a consulting firm because his boss continuously asked him to bill his time to a project that he was not working on, and when it was brought to the attention of higher ups in the firm he was told that it was standard practice.
- *Sharing confidential information about a client*. Sometimes information about a client slips out in a discussion regarding a similar project. Sometimes we inadvertently give a competitor a copy of a survey or engineering drawing, which is not a matter of public record. Assuming that the first cannon of ethics, protecting the public safety is not at stake here, this is a breach of the client's confidence at best. I believe the legal and medical professions do a better job than engineering and surveying here. There are laws today governing the release of patient information in the medical profession.
- *Being negative about a competitor*. Yes, the Supreme Court did strike rules against advertising, bidding, and competition on the basis of price cannon from codes in the 1970s, but I believe we still should practice the Golden Rule, "treat others as you would like to be treated." I recently heard a variation of it that said, "Treat others as they would like to be treated."

- *Embellishing your resume*. I've seen many resumes over the years that, when I read them, caused me to wondered how the applicant could have had so much experience and responsibility in such a short career. I've even seen applicants claiming college degrees that they don't really possess. This information is easily verified, so just don't do it!
- *Failure to report an error discovered in plans and specifications*. This dilemma is big! It generally happens once a project is under construction or the plans have been filed in the county registry. In either case, correcting the mistake is going to cost your time and money. Don't worry about embarrassment; your client will appreciate the fact that you stepped up to fix your mistake and stand behind your work. It is possible that such an ethical breach can have impact on the public safety. Remember Bill Lemesseur and the Chicago tower.
- *Accepting gifts from contractors or vendors*. This is a question of how much is too much? In some organizations and government agencies, it is even codified. Everyone has accepted a cup of coffee or lunch purchased by a contractor or samples furnished by a vendor promoting his product. I know of a government inspector who lost his job because he accepted windows from a building being demolished on his project.
- *Being involved in a project that will have significant adverse environmental or safety impact*. In the past, this may have been more of a problem but with today's environmental permitting requirements, adverse environmental impact is almost impossible. Being involved with projects that deal with unsafe products or situations is still an important issue.
- *Participating in a deceptive presentation to obtain work*. Firms have been known to promote themselves as having more expertise with a particular project type than they actually have. This can be a chicken and egg situation. How do you get experience if you don't do new types of projects? If you are faced with this situation consider hiring qualified consultants to participate on your team.
- *Failure to disclose relationships*. In some cases, professionals own or have an interest in another company or organization. If an engineer or surveyor has such interests and he/she presents a conflict in furnishing independent and unbiased services to a client, that interest should be disclosed. An example would be a surveyor who is a partner in a development that is a competitor to a new client's project. An engineer may be a stockholder in a construction company that is bidding on one of his firm's projects. While each of these may or may not be a conflict, you should disclose the relationship and let your client decide.
- *Conflict of interest*. The preceding relationship may be described as business conflicts of interest, but there are more situations where conflicts

of interest arise. A surveyor or engineer may sit on a planning or zoning board that grants approvals for his projects or he may hold public office or sit on the board of a local nonprofit organization where he has the ability to influence which firm or contractor gets a new project. Some professionals feel that in this situation they only have one vote, so their ability to affect the outcome is minor. I feel that the ethical thing to do is to make full disclosure of the potential conflict of interest to the other members of the board and to remove yourself from the discussion and voting on the issue. If this type of conflict is a regular occurrence, you should consider removing yourself from the board or resigning from office.

PROFESSIONAL VERSUS NONPROFESSIONAL ETHICS

A common definition of a professional includes having skills obtained through a highly specialized theoretical body of knowledge, which is generally obtained only through formal education. This theoretical knowledge allows the professional to formulate unique practical solutions to their clients' needs. Nonprofessionals often obtain their skills through on the job training, technical schools, or apprenticeship programs. Therefore, their skill is practically oriented toward "how to do it" rather than "what is the theoretical basis for doing this?" Application of a nonprofessional's skill is generally toward a specific technical need and not toward an in-depth analysis of the problem to be solved. Nonprofessionals often do not recognize ethical issues when they arise. They are guided more by regulations and laws than personal values, morals, or a code of ethics.

Professions generally have convinced the governing body, in the case of engineers and surveyors, the state licensing board, that their knowledge and skill is beyond the means of a layperson to comprehend and understand, and therefore they should be regulated by their peers. This self-regulation, in essence, grants the profession a monopoly to the practice skills of the profession. Ethical behavior in the public interest is necessary to keep the public confidence and maintain the monopoly. As we'll see, the public often is skeptical of professional behavior and sees self-regulation as a way of protecting our own.

Ernest Greenwood used a sociologic approach to describe professions, in his paper "Attributes of a Profession" published in *Social Works*, July 1957. It describes five attributes of a profession, which are still recognized today.

a. *A systematic theory*. Today this attribute is known as a "body of knowledge." This is mastery of systematic theory and knowledge, which is acquired through intense education, provides the background for practice. In recent years the explosion of technology and the need for management and people skills in both the engineering and surveying professions has created a movement toward even more theory and education being needed for entry into the professions.

b. *Professional authority*. Greenwood said that a professional's education imparts knowledge, which "highlights the layman's ignorance" and that the professional "dictates what is good or evil for the client since he can't diagnose his own needs or discriminate among the range of possibilities for meeting them." In addition, the client is unable to evaluate the quality of the professional service rendered without consulting another similar professional. A professional's earned authority is limited to their specific field of education and expertise, and they must not use it to exploit their client.

c. *Community sanction*. The power of the profession on its members is both formal and informal. The profession has formal power of accreditation over the colleges and universities, which offer the professional technical education in engineering and surveying. Granting or withholding accreditation controls the caliber of curriculum and instruction and in effect controls the number of schools offering professional technical education, since without an accredited degree, a professional may find it impossible to become licensed or even employed. The profession also holds the power of sanction over its members through the licensing process. Not only must a candidate for an engineering or surveying license possess an accredited degree, have specific experience, and pass examinations prior to becoming licensed, but also once obtained, one's license can be lost for violations to the licensing law of the jurisdiction. Many licensing jurisdictions also include a code of ethics in their licensing law.

d. *Codes of ethics*. Each profession develops a code of ethics, which dictates the conduct of the professional engineer and surveyor toward the public, clients, and fellow professionals. A frequently asked question today is, "is there really a need for a code of ethics?" It seems that what can or cannot be done by professions is becoming grayer and dictated more and more by laws, which require the professional to first ask if his action is legal before asking if it is ethical. In fact, most engineers and surveyors seldom consult a code of ethics to resolve a dilemma, first they refer to the law.

e. A *culture*. The professions of engineering and surveying both have formal and informal cultures. Formal cultures are organizations, such as professional firms, where one practices or offers his services. There also are professional associations or professional societies, where one receives continuing education in the field and that seek to promote the interests of the profession to the members and the public. Informal cultures could be those recognized by the public such as nonboisterous demeanor, conservative dress, and pockets full of scales, a calculator, and a variety of pencils. Surveying culture is commonly understood as one's knowledge of the land and boundaries and readiness to work in the outdoors. Surveyors often are depicted as standing behind a transit.

VALUES & MORALS

The next step in understanding ethics is understanding a professional's values and morals. A person's core values are formulated early in their life by examples set by their peers, parents, society, and siblings. This values profile is made up of characteristics that generally guide the person throughout their life both personally and professionally. It is unlikely that a person would practice a set of values in daily life and a different set in their professional practice.

Diligence—A long, steady, and persistent effort. It often includes attentive detail and care about what one is doing.

Efficiency—The personal commitment to producing efficiently and effectively.

Equality—Treating everyone the same and allowing each to enjoy equal rights. The "Golden Rule" comes to mind here.

Equity—Being just, impartial and fair in all of your dealings.

Freedom—Being and allowing freedom from restraint. The ability to practice your profession without outside influence.

Honesty—Being truthful in your discussions, deeds and actions.

Honor—Earn the esteem, respect and reputation of others.

Integrity—Keeping promises and fulfilling expectations and obligations.

Knowledge—Develop understanding gained through experience and/or study.

Loyalty—Faithfulness to one's relatives, friends and employer.

Pleasure—An enjoyable sensation, situation, things or emotion.

Safety—Freedom from danger and injury in personal, work and life situations.

Security—Freedom from doubt, minimized risk, and reliability and stability concerning the future.

Trust—Reliance on the honesty, integrity, ability, and character of a person or thing.

Stuart Walesh, author of *Engineering Your Future*, describes the following 14 value characteristics in his chapter on ethics. I believe that a clear understanding of these, as well as moral behavior, is key to ethical practice since most engineers and surveyors don't often refer to a code of ethics.

If you are familiar with Abraham Maslow's Hierarchy of Needs, you will recognize several of his characteristic "needs" in the preceding list.

Your values will be revealed in situations that arise each day, not just in ethical dilemmas.

What are morals and moral responsibility? How do they compare with values? Morals are commonly defined as our *ability* to think through the implications of our actions and how they will affect others. Therefore, they are affected by our core values. By definition, it is implicit that morals can only apply to individuals and not corporations. It is not that easy though, since corporations are made up of people. As previously discussed, Roger Boisjoly behaved as a moral individual, but his impact on the corporation was minimal. However, his recommendations were overruled by another individual, who may not have completely evaluated the moral dilemma and who allowed his commitment to the corporation to influence his decision.

Hart discusses that moral responsibility can't be attributed to corporations but says that there are four different "senses" of responsibility that can be.

a. *Role responsibility*. This goes with roles, tasks and assigned jobs.
b. *Causal responsibility*. A corporation and an individual can be found responsible for having caused something to happen.
c. *Liability responsibility*. A corporation and an individual also can be found liable and made to pay for damages.
d. *Capacity responsibility*. Primarily, that an individual can be found to have the psychological capacity for legal competence.

Hart doesn't include moral responsibility in his "senses," since he is only interested in responsibility as it relates to the law. However, moral responsibility, which can be individual or group responsibility, can be defined as:

Forward-looking responsibility. Looking for something that might or will take place. In the moral sense, it is a duty of caring and concern for what might happen.

Backward-looking responsibility. Responsibility for something that has already taken place. Looking for who is responsible for an action which already took place.

Both of these responsibilities look to the individual's causal or liability responsibility. Moral responsibility is not exclusive, as contrasted with Hart's other "senses." He uses the example that a father's responsibility for his children does not exclude the mother's responsibility. Furthermore, he says, "In the moral sense there are some things everyone is responsible for and one is safety." These can also be described as "collective responsibility" or responsibility that falls on many people at the same time.

In her book *Ethical Issues in Engineering*, Deborah G. Johnson (Prentice-Hall, 1990) states, "One of the problems of our times is that we have reduced responsibility to the four kinds described by Hart. This renders all responsibility exclusionary and provides theoretical support for wholesale abdication of moral responsibility." Society today is constantly looking for someone else to blame and to fix liability and, therefore, responsibility on.

Kenneth Alpern in his paper, "Moral Responsibility for Engineers" promotes what may be a radical thesis that:

> Engineers have a strong moral obligation, where strength of the obligation is to be understood in terms of the degree of personal sacrifice that can be demanded.

In this thesis he claims that engineers, technicians, and low-level managers are the guardians of society, and that they have a moral responsibility to protect society from harm that could result from a harmful technological development. He believes that they must be ready and willing to risk their jobs and make personal sacrifices to protect and promote the public safety and welfare.

His corollary to the thesis is the "Corollary of Proportionate Care," which states:

> When one is in a position to contribute to greater harm or when one is in a position to play a more critical part in producing harm than is another person, one must exercise greater care to avoid so doing.

His corollary states that engineers can be held to a higher standard of care, and it can be demanded that they be willing to make a greater sacrifice than others for the sake of public welfare. He goes on to say that these principles do not apply only to engineers but to anyone in a position of power that can produce harm. "If one is not willing and able to make sacrifices, then one should not seek or accept the position of power." Alpern believes that to do so would be to act immorally.

He gives little credence to excuses such as "I'd loose my job if I didn't . . ." or "If I don't do it, someone else will."

Alpern's radical thesis and corollary seem to take a position of what should happen "in an ideal world" and probably makes many engineers and surveyors uncomfortable with the level of responsibility and sacrifice (for which they don't seem to be adequately compensated) that seems to be placed upon the professional.

ETHICAL OBLIGATIONS TO OTHERS

In the professional sense, engineers and surveyors have ethical obligations to society, clients, employers and fellow professionals. I believe that whether prescribed by a code of ethics or not, these obligations flow from core values and moral responsibilities of the professional. Each of these obligations is explored in the sections that follow.

Obligation to Society

Kenneth Alpern and others believe engineers have the ultimate responsibility for the public safety and welfare and, therefore, society in general. He says engineers have the highest calling to protect the public. He believes that engineers must understand and accept this responsibility and be ready and willing to make personal sacrifices when the public safety is in jeopardy. William Wisely in 1978, stated that "the engineer shall apply his specialized knowledge and skill at all times in the public interest, with honesty, integrity and honor." He also suggested that this basic obligation or creed might serve as a replacement for a code of ethics.

Many believe that the most common dilemma of engineers in regard to the public safety is the conflict between the engineer's responsibility to society and the public safety versus his obligation of loyalty to his employer. Roger Boisjoly's concern of safety regarding the O-rings on the space shuttle Challenger and his obligation to his employer Morton Thiokol is a fine example of this dilemma. Steven Unger, in his paper "Controlling Technology: Ethics and the Responsible Engineer," outlines the increasing concern of codes of ethics for the society and public safety. The 1912 code of ethics of the American Institute of Electrical and Electronic Engineer (AIEE) "mentions the public in a patronizing manner (e.g., the engineer should endeavor to assist the public to a fair and correct general understanding of engineering matters)." The 1947 code of the Engineer's Council for Professional Development (a predecessor to the National Society of Professional Engineers) mentions "fidelity to the public" and the engineer's "duty to interest himself in the public welfare. . . ." The 1963 version of the ECPD code elevates concern for public well-being to the third of its three fundamental principles. Principle III read "will use his knowledge and skill for the advancement of human welfare." In the first canon that follows the principle states that "the engineer will have proper regard for the safety, health and welfare of the public in the performance of his professional duties." In 1974 this principle was promoted to Principle I and the first cannon strengthened to read "Engineers shall hold paramount the safety, health and welfare of the public. . . ." This provision remains unchanged in the 2007 NSPE Code of Ethics. It is obvious that concern for the public safety has gained increasing importance in the evolution of codes of ethics, but does practice really reflect this?

Obligation to Clients

An engineer and surveyor's next obligation extends to their client who must trust the professional's superior knowledge and experience. The previously mentioned conflict of interest situation applies here. An engineer or surveyor must disclose any conflict that may affect his serving as an independent and unbiased representative to his client. The engineer's obligation to his client may sometimes seem to be in conflict with his primary obligation to the public safety. What if a client's project presents an obvious environmental problem such as a high potential for ground water contamination or the illegal disposal of a small quantity of hazardous material? What about a client who requests an engineer to design a building that doesn't comply with building code requirements for snow, wind, or seismic loads? What about a client who asks that a surveyor not show a building setback on a plan because it is a violation of a zoning requirement? In these cases, the obligation to the public trumps the wishes of the client, and the engineer (surveyor) is ethically obligated to inform his client of his obligation to act in behalf of the public safety. The engineer or surveyor may ask his client to report a potential violation themselves in order not to appear that they have breached their loyalty to the client. Today, many of these situations are also covered by laws requiring the engineer or owner to act in behalf of the public safety and report violations to public agencies. Another possible conflict of the engineer's obligation to his client is the requirement, when conducting a test, analyzing data, or preparing a report, that he act as a finder of fact, whether it is to his client's benefit or not. Engineers who do forensic investigations sometimes find themselves needing to explain this ethical requirement to the client's attorney, since the code of ethics requires him to represent the interests of their client in all cases.

Obligation to Employers

An engineer's ethical obligation to the public must transcend his obligation to satisfy the needs for short-term gains demanded by his employer. Since most engineers work for corporations, this may be difficult, especially in large or publicly held corporations, where the emphasis is often on the quarterly profit/loss or earnings per share figures. Engineers who work for corporations do not have complete autonomy to evaluate a situation and act ethically regardless of the requirements of their employer. Tension constantly exists between an engineer's need for professional autonomy and the expectation of loyalty from his employer. The National Society of Professional Engineers Code of Ethics recognized this conflict and addresses it in a footnote to the Code which states:

> In regard to the question of application of the Code to corporations vis-à-vis real persons, business form or type should not negate nor influence conformance of individuals to the Code. The Code deals with professional services, which services must be performed by real persons. Real persons in turn establish and implement policies within business structures. The Code is clearly written to apply to the Engineer, and it is incumbent on members of NSPE to endeavor to live up to its provisions. This applies to all pertinent sections of the Code."

Most corporations see engineers as part scientist and part businessman. Edwin Layton, in his paper "The Revolt of the Engineers: Social Responsibility and the American Engineering Profession," suggests that American engineers have not been able to obtain the professional recognition and status they've sought because of their "deep rooted ties" to the corporation. He also states that some of the problem lies with engineering being the career path for most engineers into more highly paid management positions. Therefore, engineers who work in corporations must constantly strive to be first engineers and second businessmen, while recognizing that the most likely path for their advancement is along the lines of management not engineering. Arthur Morgan the first head of the Tennessee Valley Authority (TVA) felt that the ability of engineers to act as "free agents" was difficult to achieve, since engineers in corporations are "a technical implement of other men's purposes." He had hoped that engineering societies would keep engineers from having to act as "lone individuals," but he found that this was difficult since business influence had penetrated the "very citadel of the profession, the engineering society."

Obligation to Fellow Engineers

Although the Supreme Court struck down the provisions in the codes of ethics that forbid professional advertising, trying to supplant another engineer, or bidding against other engineers for projects, there are still common courtesies regarding how engineers and surveyors treat each other. Being truthful and extending common courtesy to fellow professionals is still practiced. Most surveyors extend professional courtesies such as exchanging plans when another surveyor is working on a project adjacent to one they are doing or of the same property if the surveyor had originally surveyed the parcel. This exchange assumes that plans are a matter of public record and their client's confidentiality is not breached. An engineering firm sometimes follows another engineering firm for follow-up projects or even continuing to work on the original project with the same client. This generally is for reasons that the client determines. In the case of a public agency, it may be a requirement to select the most qualified firm for the project or to spread the work among several firms for officials not to appear to favor one firm over another. If the work is to follow up where an engineer has been released from a project, my firm requests that the client officially sever its ties with the original engineer through written notification. As a professional courtesy, we may also follow up with the original engineer to be sure that he has been paid for his services.

Another situation arises often because of the shortage of technical professionals. A firm may be seeking to fill a position with an engineer who has the exact qualifications a firm needs but works for a competitor. This is a delicate situation and needs to be handled carefully. The firm that has the opening should advertise the position to the profession or public in general through standard advertising channels and may send an invitation to apply to the sought after person but shouldn't actively recruit him at his place of employment. If the person indicates that he is interested in the position or would "like to talk about the position," the firm should indicate that it may be best if he notify his current employer that he is considering a new position at the advertising firm. Any job offer extended to the sought-after engineer should indicate that it is expected that the engineer will give reasonable notice to his current employer, and there should be no discussion of the offer being contingent upon bringing existing clients to the new firm.

CODES OF ETHICS

Are codes of ethics commonly referred to by professionals when ethical dilemmas arise? Since most liability and responsibility are now a matter of law, and many ethical situations have been resolved by laws, are codes even necessary anymore? What do codes do? Are they even enforceable? Do professional societies or state boards of licensure ever censure professionals for ethical violations? Based on preceding discussions, an argument may be made for the necessity of codes of ethics based on the application of core values and an engineer or surveyor's moral responsibility, which is very difficult to put into law. Codes of ethics are seen as a higher professional requirement than the law alone. Why then is ethics not taught with more rigor in engineering and surveying curriculums and why aren't ethical considerations discussed more in daily practice? Let's review where codes of ethics came from, their reason for being, and their application to today's practice.

It appears as though the original professional code of ethics arose at the Royal Society of Physicians in the 1800s. As previously mentioned, the AIEE had a code of ethics in 1912, so codes been around for a long time. Establishment of a code of ethical practice is one of the first things that professional societies do to differentiate themselves from nonprofessionals. Establishment of such codes also attempt to convince the public that the profession is truthful and fair in its dealings and worthy of self-governance and therefore a monopoly on practice established through the state licensing law.

What is the purpose of a code of ethics? According to McCuen and Wallace, there are four:

a. Establish goals of the profession.
b. Identify values.
c. Identify the rights and responsibilities of the organization and its members.
d. Provide guidance and inspiration.

While these may have been the original purpose for establishing codes of ethics let's critically examine whether or not they still apply. Since most engineering society codes of ethics are very similar to that of the National Society of Professional Engineers (NSPE), we'll use it for this analysis. A similar analysis can be with The National Society of Professional Surveyors Creed and Cannons.

Goals of the Profession

Professional societies were formed by groups of professionals seeking recognition, status and to clarify their goals and values along with those of fellow professionals. Presumably, codes of ethics were developed and adopted to govern the behavior of the society's members. In today's interactive technological world, several engineering and related professionals may work together regularly, but their professional goals may be quite different. While a civil engineer, because of his involvement with the public infrastructure (and manager of the entire project), may be focused on the public safety in every phase of his work, the mechanical engineer, whose narrow focus is the pumps and piping in the water treatment plant, may see a goal of his profession to provide the most energy efficient systems to minimize maintenance and operating cost. The impact on public safety may be far from his mind. Similarly, electrical and computer engineers may see their profession's primary goal to be the cutting edge of developing a faster processing chip, a better operating system, or a more integrated communication system. Only the civil engineer may even be aware that the first fundamental cannon (and therefore the primary goal of the project) is to "Hold paramount the safety, health, and welfare of the public." The other engineers may not even be members (or see the need for membership) in a professional society.

Values

Many engineers and surveyors profess core values similar to those previously discussed yet see no need to be members or participate in a professional society. This seems to be the trend of younger professionals who are busy as part of dual-profession families who may be raising children. Does this lack of involvement or even lack of licensure make them less ethical than those who worked hard to establish the profession before them? In most cases ethical considerations of their practice are far from day-to-day occurrences. Without even knowing it, they may be following the fourth fundamental cannon, "Act for each employer or client as a faithful agent or trustee," or the sixth cannon, "Conduct themselves honorably, responsibly, ethically and lawfully so as to enhance the honor, reputation, and usefulness of the profession." This is the case with most young professionals that I'm associated with. They all seem to have good core values and practice moral responsibility.

Identify the Rights and Responsibilities of the Organization and Its Members

Why is it that many engineers and surveyors are not involved in their professional organizations? Some may argue that they have no relevance and "don't do anything for me." Much is written in professional association publications about the organizations lack of improvement of the stature and financial well being of its members. Many engineers employed in large organizations are members of labor unions. Is this due to a requirement for employment or do unions do more than professional organizations to enhance the rights of their members? Such membership probably has little to do with ethical practice.

Provide Guidance and Inspiration

Codes of ethics even when read to seek guidance regarding a particular ethical dilemma are boring reading and, in my opinion, not very inspirational. While inspiring professionals toward ethical practice (and the constant updates and revisions) may have been an original goal, most engineers seek guidance from the law and fellow engineers first and only rely on the code of ethics as a last resort. It is unlikely that those who see little need for professional society membership or a code of ethics would seek clarification of the second cannon, "Perform services only in areas of their competence," or the third, "Issue public statements only in an objective and truthful manner."

ENFORCEABILITY OF CODES OF ETHICS

If codes of ethics are to be meaningful (and there is some question as to whether they really are) how are they to be enforced? Professional societies have been rather weak for a variety of reasons, not the least of which has been the reluctance of individual members to bring ethical complaints against fellow members. Some complaints and requests for censure may, however, arise. These seem to be best dealt with by state licensure boards who have adopted a code of ethics as part of their rules for practice.

Guidance for Solutions to Ethical Dilemmas

Whether engineers or surveyors are members of professional organizations or not, if we believe that our profession has a higher calling to protect the public and our clients, and show loyalty to our employers and fellow practitioner, we eventually will be faced with an ethical dilemma. Where do we turn for guidance? The case study method is a recognized way of seeking guidance and possibly a solution. The National Society of Professional Engineers, www.nspe.org, has been publishing cases that analyze ethical dilemmas since 1976. Educational institutions like the Murdough Center at Texas Tech College of Engineering, www.niee.org, present a library of case studies for review.

Professor Robert Tillman of Northeastern University presents the following 10-point guideline for deriving solutions to ethical dilemmas.

1. Determine the facts in the situation—obtain all of the unbiased facts possible.
2. Define the stakeholders—those with vested interest in the outcome.
3. Assess the motives of the stakeholders—it is important to understand what influences individual stakeholders to act the way they do in a specific situation including hidden agendas or business practices.
4. Formulate alternative solutions.
5. Evaluate proposed alternatives—look for a win-win solution.
6. Evaluate consequences of decisions on all stakeholders.

7. Seek appropriate additional assistance/input—including the advice of peers and supervisors.
8. Select the best course of action.
9. Implement the selected solution—take action as warranted.
10. Evaluate the decision.

THE ROLE OF MENTORING

I've discussed the fact that many, even most, young professionals do not belong to either a professional society or are licensed and most don't know where to find a code of ethics to consult when an ethical dilemma arises.

Since many moral and ethical dilemmas are resolved today by law, regulations, or corporate decision making processes or in some sort liability/responsibility forum, (i.e., courts, arbitration, mediation) are codes of ethics are still relevant? What is the role of a senior-level mentor to help educate young professionals in the topic of ethics? I believe that most young engineers and surveyors enter the profession because they believe that not only are they pursuing an enjoyable, challenging, and satisfying way of life but that they are providing some sort of public service. Even though the topic of ethics is not well covered in most engineering and surveying curriculum today, I feel that most, if not all, graduates leave university with at least this basic level of commitment to the public good. It becomes the responsibility of those of us with many years of experience to mentor them.

The first task is to elevate their level of commitment to the public in general to one of commitment to "public safety, health, and welfare." Young engineers and surveyors receive a good technical education, but upon graduation are really "technicians" with very little understanding of what it takes to become a "professional" or what ethical issue may face them in their career. As mentors we must impart to them Alpern's belief that they have a "higher calling" than just solving equations, running computers, and arriving at the best technical solution. This is best done by involving them with client relations and the public involvement process early in their careers. Ethical dilemmas include people and if young professionals don't meet clients and understand the public process, they will have little appreciation for the impact of their technical work.

Second, mentors must make it clear that in order to be a professional, effort may be required outside of the ordinary 8–5 day. Other professions have done a better job of this than engineering and surveying. Young lawyers understand that long nights and weekends may be necessary to adequately prepare for a case that will be coming to trial. Young emergency room doctors certainly understand that they just don't leave the emergency room at the given shift departure time if a crisis is being treated. Accountants understand that serving clients requires a commitment of almost 24/7 as April 15th draws near. Most of our young surveyors understand that sometimes it only makes sense to work a longer day to finish an assignment rather than spend additional travel time to return to a distant project for an hours worth of work. In some cases, our firm requires young engineers to attend evening meetings where their client's projects are being considered. Sometimes longer days and "extra curricular" meetings present conflicts with personal commitments, family, and friends. Commitment is part of being a professional.

Third, mentors should set an example by demonstrating high core values and moral responsibility to all in both personal and professional life. Ethical professional practice is based on a professional's reputation for having high values and doing the right thing. Mentors must remind young professionals that in the "Is it legal? versus Is it ethical?" realm, the higher calling of ethical responsibility prevails. The Golden Rule comes to mind here.

Finally, a mentor should emphasize and promote professional licensing. Many young professionals never pursue licensing because they work in an industry where it is not required. These young professionals will argue that they have as considerable client contact and high levels of technical expertise and responsibility, and their products have just as great an impact on the public safety (their companies can be held liable through consumer protection laws) as those who are licensed. While this argument may be valid, other than high core values and moral responsibility to ethical behavior, they seldom are called upon to demonstrate Alpern's highest level of ethical responsibility by putting their job on the line (possibly through "whistleblowing" as an extreme example) if necessary. Employed in a large corporation, it is easy for them to say that their design (possibly an unethical one) is only carrying out corporate mandate. They haven't made an ethical commitment through obtaining a license and, therefore, have little to lose (possibly their career) if they engage in unethical actions. Solving ethical dilemmas may have little to do with most of their work.

Seniors firm members come from a different time and background. They've faced many ethical dilemmas over the years, possibly some now resolved by new rules and laws. This experience should be shared with young professionals so that they understand that the issue is not just whether is ethical versus legal, but really believe that they have the highest calling for protection of the public safety, health, and welfare.

7

BASICS OF FINANCIAL ACCOUNTING & ANALYSIS FOR MANAGERS

INTRODUCTION

Why is it necessary to understand accounting and a firm's financial information? Can't we just leave this financial stuff to the accountants? The reason is similar to learning to read a road map. If you don't understand where you've been, how can you figure our where you are going? Some small surveying and engineering firms manage their finances by the checkbook method. That is, as long as there is a positive cash balance in the account they feel they're making money. The principal's interest is in the technical aspects of the profession and maximizing their billable time. Finances are left to the bookkeeper and the accountant. There are several problems with this method. The most important one is that cash in the checking account doesn't necessarily equate to billings or profit. Other financial items such as accounts payable (the money owed to others) and accounts receivable (the money owed to you) are not reflected in the checkbook balance. Neither are the physical assets of the company. The checkbook also provides little information as to assets or liabilities, how income and expenses compare to last year or how your firm is doing compared your competitors or to the industry as a whole. Accounting for a professional practice is not the same as a small manufacturing business or a retail shop. Be sure to engage an accountant who understands the difference.

Many small firms understand that "cash is king" and without it you will need to borrow in order to sustain operations. Some say that the best way to weather a downturn in the economy is to "pay your bills and accumulate cash." Both of these principles are sound financial practices but neither provides much insight into understanding the overall financial health or future of your firm.

In this chapter, you'll learn how to understand the two most important financial statements, the balance sheet and income statement as well as how to manage cash flow, and measure profit. We'll discuss the difference between cash and accrual accounting. You'll learn what overhead is and how it can have a significant effect on doing business with some types of clients. In Chapter 8 we'll analyze the overall financial health of your firm. We'll discuss how to measure your firm's health using key ratios and to compare how your firm is doing when compared with the industry. You also will learn how to price projects and how to set your fees by taking into consideration salary, overhead, and expected profit as well as how to charge fees based on "value added." Methods for determining the work backlog and how to determine your breakeven point will also be discussed.

The Profit Motive

Many small surveying and engineering professionals feel self-conscious about charging an appropriate fee and fail to see the real value of their work. I believe this comes from the belief that we are a low-paying profession with a low level of self-esteem and understanding of our position as it relates to other professions. We feel that we get little respect from the public, and that many clients don't see the value of our work. We often feel that our hourly rates are too high, and the client doesn't understand the level of effort (man-hours) behind the work product we produce, so we arbitrarily reduce our bills to "something reasonable." When was the last time your auto mechanic reduced his invoice because you were shocked at the amount it cost for a routine tune-up or when was the last time you got a free haircut because your hair dresser or barber thought that making $120 per hour was too much to earn for their level of training and technical expertise?

We'll discuss how to get clients accustom to paying their bills in a timely way and how to track payment through aging accounts receivable. Sometimes it is necessary to decide when to write-off an old bill or try and collect it through a collection process.

Profit is *not* a dirty word. This doesn't mean that we should cut corners or sacrifice quality to earn a reasonable profit. The quality of our work product should never be compromised. But, if we are consistently over budget, and it always takes longer than we think it should to accomplish a task, we need to examine our processes and procedures. Careful analysis and reexamination of our workflow may reveal that considerable rework is being done or stop-start procedures, which are inefficient, are taking place. Do we have staff that work aimlessly without a clear understanding of the budget or goals?

They may be the reason for budget busting. Are we providing the client with the work product level of quality that they need or are we striving for a level of perfection that is unnecessary? A basic understanding of accounting and what the "numbers" tell us is a key to controlling these issues.

Profit is needed to help our firms grow, to provide cash for capital purchases, to compensate owners for risk and to pay bonuses to deserving staff. Stop thinking about being a poor low-paying professional and focus on value, quality, and profit. *Remember, profit is not a dirty word!*

How Do Small Firms Go Broke?

There are many ways for small firms to go out of business or at least perform poorly for a period of years. Some are:

- *Beginning under capitalized.* Most small firms start on a shoestring with computers and surveying equipment purchased from personal savings and very little start-up cash. Attempts are often made to keep overhead low by working out of a home office and hiring part time staff. Often there is no business financial plan, and the opportunity for success is hopeful optimism at best. If there is plenty of work, the firm may survive, but an economic downturn means almost certain disaster. I know, my firm started this way in 1974, and it was only sheer determination and a loan from a relative that pulled us through.
- *Not getting enough work.* This is the classic professional business cycle, which many believe can't be changed. When small firms are very busy, most of the effort is spent doing the work at hand with little thought or vision toward what will replace the work and keep the staff busy once the projects are complete. Since it often takes years to develop new clients and new projects, marketing efforts must be consistent no matter what the present workload. Diversification also helps counter the effects of an economic downturn. Our firm has been in business for over 30 years, and I have seen several severe economic downturns. I believe that our mix of private and public clients has been one of the keys to weathering market downturns. Just like the stock market, a diversified portfolio has the best chance for long-term success. When one market sector is down, another is likely to be up. Stay diversified and have a consistent marketing program.
- *Not charging enough to cover costs.* Some small firms believe they are profitable because they are growing and there's lots of cash in the checking account. What actually may be happening is that growth is providing the cash to finance losing projects. This is fine during a growing

economy but can produce a real financial crisis during a downturn. This may be particularly true if the firm charges clients a retainer fee and the cash is accumulated in the check account before the work is done.

- *Not controlling overhead costs*. Since the cost of doing business, rent, utilities, insurance, and benefits are our biggest expenses, we need to know how much they are and how they change on a regular basis, more frequently than annually. We need to recognize when costs should be allocated to a project as a reimbursable cost rather than flow to general or firm overhead. Some government clients limit overhead for firms that do government work. These firms may have higher than allowable overhead and may be consistently losing money.
- *Not collecting money due*. Once they have been invoiced, we need to convince clients to pay promptly. Most bookkeepers either have a running record of aged accounts receivable or can readily prepare one. Most businesses extend credit to good customers up to 30 days. Since, in reality, this may be up to 60 days after the goods are purchased on credit (you buy a widget on the first of the month, receive the statement after the 30th of the month and have to the 30th of the next month to pay) they begin to get concerned after 30 days of nonpayment. It is not uncommon in small surveying and engineering firms to have large receivables over 90 days with no plan to collect them. The principals may not even be aware of the amount of long overdue receivables. The average collection period for the surveying and engineering industry is about 67 days. You've earned this money, and it can create real problems if you don't collect it.

There are many other operational ways to go out of business. Most small firms do not fail for lack of knowledge of the industry or technical expertise; they fail due to lack of planning and financial understanding and control by the principals of the firm.

ACCOUNTING BASICS

Generally Accepted Accounting Principles

There are several generally accepted accounting principles that need to be understood before we can get into the actual details of understanding financial statements. These are taken from *Accounting and Financial Fundamentals for Non-financial Executives*, by Robert Rachlin and Allen Sweeny.

- *The money measurement principle*. This principle simply states that accounting transactions are only measured in terms of money. This provides for consistency and allows all firms to be compared on an equal basis. It also presents some problems, since intangibles such as the value of goodwill and reputation created by years of a principal's hard work or the value of most firm's greatest asset, their employees, is not accounted for on the financial statements.

- *The business entity principle.* This principle simply says that accounting is done for business entities and not individuals. We found out about this principle the hard way. When our firm was founded we were a proprietorship and even though we had a separate company checking account, it was difficult to separate our personal financial lives from the business. Annual profits and/or losses caused our personal tax situation to be a "yo-yo," and we had to pledge our personal assets in order to borrow money to purchase new company equipment or vehicles. A word of advice, set up a separate business entity, if you really want to know how your business is doing.
- *The going concern principle.* This is the idea that a business is intended to exist for a long period of time. It is necessary so that accountants can account for things like depreciation and long-term debit. It is closely tied to the following principle of cost.
- *The cost concept principle.* Most accountants prefer to value assets on the basis of their original costs. This simplifies things greatly, since the only thing needed to establish the value of the asset is the original invoice or statement. It also simplifies depreciating the value of the asset. The real economic value of an asset may be considerably different than the cost basis carried on company's books. For example, real estate and office buildings often appreciate rather than depreciate, so although they may have a very low asset value on the balance sheet, they are worth considerably more on the open market.
- *The realization principle.* This principle says that the revenue (income) is realized at the time the goods or services are delivered. In the case of an engineering or surveying firm, this may mean when the drawings and specifications are delivered or the survey plan is complete and delivered to the landowner. This isn't necessarily the same as the date of the invoice for the services or the date of receipt of payments. In fact, most of the time it isn't, but the realization principle says that this is when you record the income. This is the basis of accrual accounting, an explanation of which follows.

The Time Accounting Model

Surveying and engineering firms only have one product to sell—time. If you have, and effectively use, a time accounting model with a corresponding integrated software package, you have the basis for financial success. Many years ago my MBA thesis was to evaluate time accounting software and select the one that I thought would produce the best reports and results for our firm. I can honestly say that this was one of the most important business decisions our firm has ever made. Those were the days of infancy in such software for architecture, engineering, and surveying firms. Many packages were really beta models and wouldn't do what they claimed. Large firms had in-house-developed software, and some small firms used outside services connected via dial-up modems. Most small firms kept handwritten time sheets, manually posted time to job logs, and hand typed individual invoices. Accounts payable and accounts receivable also were tracked by an internally developed manual method. We have come a long way in 30 years. Individual, fully integrated time entry accounting and project management software is affordable by all firms. The problem can be setting it up to replicate your workflow and produce the reports and invoices you need.

Time accounting starts with the individual employee's time sheet. It needs to be simple in format and function so that the staff fills it out regularly and it accurately depicts the time, phases and tasks performed on a project. Most software also provides an opportunity to insert a comment after a time entry. This detail helps when clients question what work was being done and charged to their project. It almost goes without saying that management must insist on staff regularly filling out time reports and not tolerating those who don't.

The integrated time accounting system must also include a way for project-related expenses such as mileage, printing costs, consultant fees, and other reimbursable cost to be entered and flow automatically to the project reports and invoices.

The outputs of this system, on the project side, should be project budget reports, work-in-process reports, invoices, and statements of outstanding accounts. An income statement, balance sheet, accounts payable, and aged accounts receivable should be outputs on the firm accounting side.

Fee Schedules

As I mentioned previously, our only saleable product is our time. If we don't set our fees to capture all of our costs and a reasonable profit, there is no way the firm can be successful. In many cases, we contract in different ways with different clients; therefore, the task of setting fees is not a simple as one may expect, so let's explore it.

What are the different ways we can charge for our work? The basic method is hours worked on a project times a billable rate from your fee schedule, which is published and available to prospects and clients. Typical fee schedules contain billable hourly fees for certain positions in the firm such as "principal," "draftsperson," or "survey technician." Although fee schedules tend to be a standard to charge most clients, some clients may negotiate a discounted fee schedule based on their volume of work and prompt payment. A variation of the hourly fee schedule, which is often used by government agencies is base hourly wage times an acceptable overhead multiplier plus a negotiated profit. This amounts to a fixed hourly billing rate for every staff person in the firm rather than a billing rate by job category. A word of caution is necessary here. Government agencies can be tough negotiators, who have maximum allowable overhead multipliers and maximum allowable profits. It can be very difficult for firms to make significant profits by just doing government work. Determining your actual overhead multiplier will be discussed later.

Most clients will not accept a proposal that just states that they will be billed for the hours worked times the fee schedule rate. They need a budget amount or a lump sum or fixed fee. The difference here is important. It is an issue of risk. Do you want the client to take the risk, thereby offering you relative security that you'll be paid for all of your work and earn a reasonable profit or would you be willing to assume some risk with the potential for earning a large profit or taking a loss? Hourly agreements with approved budgets are a very safe way of consistently making a reasonable profit, but given the inevitable write-offs (hours that can't be billed for what ever reason), there is no way of improving on your projected profit. You can only loose, especially if you agree that your charges won't exceed the budget. If you are willing to take some risk, you can significantly improve your firm's overall profit picture. A lump sum fee is the way to do this. Our firm often proposes lump sum fees for project types where we've got a significant amount of experience and the scope of work is well defined. Once we've negotiated a lump sum fee, we work very hard to complete it in less than the time estimated, thereby producing a greater than expected profit. Of course, the converse is true also. If it takes us longer than anticipated we loose. Lump sum fees take conscientious project management, understanding of productivity, and keeping track of scope creep, since unanticipated work will almost certainly cause losses. It also can produce significant bonuses, which our staff enjoys.

A spreadsheet should always be used to determine fees for a project since "what if" scenarios are always part of the matrix of determine scope and fee.

Cash versus Accrual Accounting

The concept of accrual accounting was introduced in the realization principle described earlier. It's a very powerful tool, allowing us to understand whether our firms are making money or not. For example, if we are conducting our business by the cash or checking account method, we may work on a very large project for a period of a month or more before we have a deliverable which allows us to invoice for our work. During that period cash will be flowing out of the account to cover payroll and other expenses, while little or no money is flowing into the account. An examination of the checkbook balance at the beginning and end of the month will show a declining balance since no cash income was received. The first assumption by a principal may be that "we haven't made any money this month," while in reality the staff has been gainfully employed and productive, but the invoice for the work hasn't been sent because a particular milestone hasn't been met. The situation can be further complicated when the following month cash is received for the previously completed project and a large checking account balance exists but no productive work is being done by staff, since the principals have not marketed and brought in sufficient work to replace that which has been completed. On the accrual basis, the firm made money in the first month and lost in the second month, exactly opposite of what the cash basis and the checking account are reporting. Now, can you see why accrual accounting is so important in order to really understand the financial condition of your firm?

Dual-Entry Bookkeeping

The preceding example illustrates the effect of accrual accounting on the income statement or profit and loss statement of a firm, but it also impacts the assets, liabilities, and owner's equity of the firm shown on the balance sheet. This is the dual-entry system of bookkeeping. The impact on assets must be accompanied by a corresponding impact on liabilities and owner's equity. It is best explained by a simple example showing how assets and liabilities offset each other and their effect on the income statement of a small surveying or engineering firm.

John Doing decides to go into the surveying and engineering business doing business as ACME Surveying & Engineering on January 1, 200x. He uses $20,000 of his own money and borrowed or leased equipment, since he can't afford to purchase any just yet. The balance sheet for his new company ACME Surveying & Engineering is shown in Table 7.1 and looks like this at start-up January 1st.

As previously mentioned, John Doing and the ACME staff worked for a month accruing billable work, but they haven't been able to bill it because of not meeting a client's specified milestone. This accrued, unbilled work

Table 7.1 ACME Surveying & Engineering Balance Sheet, January 1, 200X

Assets		Liabilities	
Cash	$20,000	Owed to Others	$ 0
Other Assets	$ 0	Owner's Equity	$20,000
Total Assets	$20,000	Total Liabilities & OE	$20,000

Table 7.2 ACME Surveying & Engineering Balance Sheet, January 31, 200X

Assets		Liabilities	
Cash	$12,000	Accounts Payable	$ 1,000
W-I-P	$10,000	Owner's Equity	$21,000
Total Assets	$22,000	Total Liabilities &OE	$22,000

represents real income and is known as work-in-process or W-I-P in most accounting systems. Let's assume that ACME paid $8000 in payroll and other expenses and accrued another $1000 in bills, which haven't been paid. At the end of January, ACME's balance sheet shown in Table 7.2 looks like this.

By the rule of dual-entry accounting the impact on assets caused by the introduction of W-I-P as an asset needs an offsetting amount on the liability and equity side of the balance sheet. As you can see, ACME went from having no liabilities at start-up to owing $1000 to creditors, but this still leaves the balance sheet unbalanced by $1000. An increase of $1000 in owner's equity seems to be the answer. Let's look at ACME's corresponding simplified income or profit and loss statement in Table 7.3.

As shown in the preceding balance sheet and following income statement, even though no cash has been received by ACME, it has made a $1000 profit for the month of January. It makes sense that the profit should be reflected in the balance sheet by increasing owner's equity, since the only additional liability incurred is the account payable $1000. This example illustrates the principle of dual-entry accounting.

If ACME had only kept track of its books on a cash basis, with no cash income during January, it would have shown a loss of $8000 for payroll and other expenses paid. The cash basis has no way of handling the $1000 in bills owed but not paid. On the balance sheet owner's equity would have been reduced by the same amount and would have been shown as $12,000 rather than $21,000, which truly reflects the January effort and subsequent profit.

Table 7.3 ACME Surveying & Engineering, Income Statement January 31, 200X

Income	
Cash for services rendered	$ 0
W-I-P unbilled	$10,000
Total gross income	$10,000
Expenses	
Accrued payroll expense	$ 6,000
Operating expenses paid	$ 2,000
Operating expenses accrued & payable	$ 1,000
Total Expenses	$ 9,000
Total net income	$ 1,000

A more detailed discussion of both the balance sheet and the income statement follows.

INCOME & INCOME STATEMENTS

As shown in the preceding example, both accrued income and expense are shown on the income statement to determine whether a firm has realized a profit or loss for the reporting period. Usually, a firm's accounting staff produces statements monthly, which are reviewed and adjusted quarterly or annually by an outside accountant. The year-end statements typically include adjustments by the accountant such as accumulated depreciation (a reduction in the value of physical assets for wear and tear), write-offs for unbilled or uncollectible accounts, and an allowance for questionable receivables over 90 or 120 days.

If your firm is just starting up or you are considering a change to the accrual system consider using a standard chart of accounts developed for the surveying and engineering service industry. ACEC (Deltex) has such a chart. Also, select an accountant who is experienced in the industry, since accounting in our industry is significantly different from manufacturing or retail businesses.

Classifying income and expenses is the next step in setting up your accounting system. Since this varies from firm to firm and may be dictated by the types of clients that you serve give this some detailed thought.

Income

As we discussed previously, income is realized when the work is done (or in some cases when it is invoiced) rather than when the cash is received. It also should be broken down in to fee income earned by the efforts of your staff or that earned by reimbursement for project-related expenses such as testing laboratories and consultants. By the way, most firms' mark up reimbursable expenses to cover their overhead cost of handling the billing and extending credit to clients. Some government clients do not allow this. If this is the case with any client, have them contract directly with the provider of the service and have the bills sent directly to the client if possible. Another income breakout item might be interest income. This income technically is derived from your firm lending money for your services to your clients and wouldn't have been derived if the client had paid their bill in a timely manner. Larger firms, which treat departments as profit centers, often break out income by department. In order to determine which client type or market sector is the most profitable, you may want to consider breaking income down in this manner, too.

Uniform Chart of Accounts

1000 – Assets

1100 Current Assets – Cash or those items convertible to cash within one year.

1100 – Cash – cash on hand or in bank accounts

1120 – Accounts Receivable – billings owed by clients

1130 – Notes Receivable – outstanding note owed by clients or others

1140 – Earned Unbilled Revenue – commonly known as Work-in-Process

1150 – Investments – investments owned by the firm

1160 – Prepaid Expenses – expenses paid in advance

1200 Fixed Assets – Assets with life loner than one year

1200 – Office equipment – furniture and equipment to be depreciated

1202 – Accumulated depreciation – depreciation charged against 1200

1210 – Technical equipment – items such as lab and survey equipment

1212 – Accumulated depreciation – depreciation charged against 1210

1220 – Transportation equipment – generally vehicles use in business

1220 – Accumulated depreciation – depreciation charged against 1220

1240 – Leasehold improvements – improvements to leased property

1242 – Accumulated depreciation – depreciation charged against 1240

1280 – Other assets – other assets not categorized above

2000 – Liabilities

2100 Current Liabilities – Liabilities due within one year

2100 – Accrued Salaries and Payroll – salaries owed but not yet paid

2113 – Accrued employer salary taxes – employer portion of payroll taxes and insurance not yet paid

2130 – Accrued payroll deductions – employee portion of payroll taxes and insurance withheld and not yet paid

2140 – Notes payable (short term) – portion of notes due within one year

2150 – Accrued Business Taxes – includes deferred Federal and state income or sales tax accrued but not paid

2160 – Accounts Payable – all amounts due suppliers and vendors including rent, consultants, etc. not yet paid

(*Continued*)

2170 – Accrues Interest and Bank Charges – interest and bank expenses not yet paid

2200 Long Term Liabilities – Liabilities due beyond one year

2200 Mortgages Payable – Portion of mortgage payable beyond one year
2210 – Notes Payable – amount of notes payable after one year

3100 – Net Worth

3100 – Capital Stock – The amount originally invested when the owner(s) started the firm.

3300 – Retained Earnings – The cumulative earnings or losses since the firm began.

3310 – Profit and Loss – Net earnings or loss for the current year.

4000 – Income

4100 – Professional fees – The amount earned on projects from clients includes overhead and profit.

4200 – Reimbursable income – earnings from reimbursable expenses such as consultants, travel, or lab fees.

4400 – Interest and dividends – income earned from company investments

4600 – Sale of Capital Assets – earnings or losses incurred as a result of sale of capital equipment.

4700 – Other income – incomes earned from other sources not mentioned above

4800 – Losses on Uncollectable Accounts – losses from write-offs and bad debts that can not be invoiced for whatever reason.

4900 – Work-in-Progress – professional fees and reimbursables earned but not invoiced.

5000 – Direct Project Expenses

5100 – Direct project payroll – payroll expenses for time expended on technical work.

5200 – Direct project expenses – expenses directly attributable to a project such as consultants, fees, lab services, reproduction fees, travel etc.

6000 – Indirect Costs

6100 – Indirect (payroll connected) costs – bonuses, vacations, leave, employer salary taxes including Federal and state, health insurance, employee education.

6200 – Indirect (General and Administrative) Costs

6210 – Indirect (nonproject) payroll all administrative and clerical salaries, time expended on nontechnical work, general supervision, business development time.

6220 – Business taxes – Federal and state income taxes in connection with the business but not payroll taxes.

6230 – Legal and accounting – charges for outside legal and accounting.

6240 – Interest and Bank Charges – bank charges and interest expenses.

6250 – Rent, Utilities and Maintenance – all costs for office space including cleaning, repairs and maintenance.

6260 – Office Supplies and Services – all costs for supplies and services not charged to projects.

6270 – Telephone – local and long distance services.

6280 – Professional Activities – includes dues, subscriptions, memberships, workshops and seminars, professional licenses etc.

6290 – Vehicle Expenses – all transportation charges not record to a project.

6300 – Business Development – all expenses in connection with getting new work.

6400 – Depreciation expense – depreciation for capital purchases deducted over their life.

Adapted from *Financial Management and Project Control for Consulting Engineers*, ACEC Guidelines to Practice, Volume II, No. 1, by Lowell V. Getz, CPA, no date.

Expenses

Your accounting system also should break down expenses for the purpose of analysis. Expenses can be either overhead or general administrative expenses, or project-related expenses. They can be broken down further to fixed and variable expenses. I suggest this be done for the simple reason that if times become tough your greatest opportunity for managing expenses lies with variable expenses since, by definition, fixed expenses remain consistent for long period or at best are difficult to change. Most of an engineering and surveying firm's expenses are fixed.

Since there isn't much you can do about them, let's briefly discuss fixed expenses first. They include most general administrative expenses: the rent, depreciation, heat, lights and other utilities, and insurance. The cost of maintaining professional licenses also is a necessary fixed expense if you are to continue to produce work and stay in business. Yes, each of these can be adjusted to a certain extent, but they are not related to the activity of the firm like variable expenses.

Variable expenses in our industry are also known as direct costs. They are the expenses that increase or decrease with the level of work the firm is doing or the level of growth or decline. The obvious variable expense is the direct labor cost. In order to properly determine the cost of doing work, or your overhead factor, labor expense needs to be broken down into direct billable labor and nonbillable labor. Direct billable labor is the cost of producing projects and generally is subtracted directly from fee income to produce gross income. Nonbillable labor becomes part of general administrative expenses and therefore part of overhead. Most technical employees have billable and nonbillable time, which needs to be tracked since an hour spent not producing billable income produces a increase in general administrative costs. Administrative staff who are completely nonbillable are more of a fixed expense than a variable expense, since their work is not is only partly related to the firm's level of work. Other variable expenses could be bonuses, the cost of seminars or continuing education, professional society memberships, subscriptions to periodicals or journals, charitable contributions, pro bono work, income taxes, and office supplies.

Project-related expenses include reimbursable expenses that are directly billable to a project and should not be considered general administrative or overhead expenses. They are variable expenses. These include outside travel expenses, laboratory fees, and special consultants. Some firms that need to strictly control their overhead record project-related telephone calls, copies, travel, survey bounds, and computer time as project reimbursable expenses. Unless a good system is developed to do this, it may be more administrative effort than the return produces.

Income Statement

An income statement reports the financial health of a firm over a specific period of time, usually a month, quarter or year. In small companies, the income statement is important in determining how well a company is doing from month to month or from year to year. It reflects growth or the lack thereof. Assuming that by now you are convinced that your income statement should be prepared on an accrual basis, net profit also has an impact on the balance sheet, since net profit produces an increase in owner's equity and a loss reduces it. The revenue or gross income side of the income statement is not the only thing affected by accrual accounting. The expense side contains accrual items, too. These are accrued expenses unpaid, depreciation, adjustments for uncollectible fees (previously accrued as income), and anticipated or accrued income taxes. Each of these has the effect of reducing income, but again it is a truer statement of a firm's financial condition than cash-only statements or just checking the checkbook balance. Some firms distribute large annual expenses such as professional liability insurance as monthly accrued expenses in order to evenly distribute the expense over a year. Another expense that reduces net income or profit is bonuses. This is definitely a variable expense, since if a firm isn't making money there is very little money available for bonuses. It also is used as a method for controlling taxes, since most firms would rather pay out excess income to the owners and staff than to the government. Technically, bonuses are called "discretionary expenses." This means that bonuses can be declared and accrued at the discretion of the firm's principals. When real profit is analyzed, such as to determine the value of a firm for sale, the discretionary expenses are added back to produce a "real" profit value.

When analyzing an income statement, compare it with the previous period whether a month, quarter, or year. Look for trends: *increasing* nonbillable labor cost and *decreasing* billable labor cost. Compare total labor cost with that of past periods. Also look for unusual increases in specific expense items such as insurance and utilities. If they are variable expenses, there may be a good explanation but question increases or decreases in fixed expenses such as rent or depreciation. (See Table 7.4.)

THE BALANCE SHEET—A SNAPSHOT IN TIME

In today's world, owners of companies, both public and private, seem to be focused on the bottom line, profits. There is nothing wrong with focusing on the income statement and profits (I emphasized the need for profits previously), but it may be a short-term perspective on a company. In times gone by, there was more emphasis on the balance sheet, assets, liabilities, and owner's equity or "what you have in the bank." This was because of the longer-term outlook on a company and the fact that a month, or even a year, of losses may be able to be weathered if a firm has sufficient assets, in the form cash reserves, and the impact on the owners, in terms of reduce owner's equity, could be accepted. In order to completely understand the financial health of a firm, both financial statements must be used together. They are very different. The income statement can be analyzed for trends in income and expenses that occur over a period of time, such as a month. The balance sheet is constantly

Table 7.4 ACME Surveying & Engineering, Inc.
Income Statement January 1 to December 31, 20XX
(Accrual - figures rounded to $1000)

Revenue	
Project Fee Income	$2,100
Other income	$ 15
Consultant & Reimbursable Expense	$(250)
Total Net Revenue	$1,865
Expenses	
Direct labor (billable)	$ 600
Indirect labor (admin & nonbillable)	$ 250
Leave	$ 90
Payroll expenses - FICA	$ 80
Payroll expense - health insurance & HAS	$ 48
Payroll expense - state taxes	$ 3
Rent	$ 60
Office & technical supplies, postage & repairs	$ 75
Insurance - general & auto	$ 15
Insurance - professional liability	$ 30
Cleaning & maintenance	$ 25
Marketing, advertising & business dev	$ 22
Utilities (heat, elec, telephone, water)	$ 20
Auto expense, repairs & registration	$ 10
Legal & accounting	$ 5
Licenses, dues & subscriptions	$ 2
Depreciation	$ 10
Interest	$ 5
Contributions	$ 10
Taxes (NH BET)	$ 3
Employee education & seminars	$ 20
Bad debts	$ 20
Miscellaneous expenses	$ 5
Total expenses	$1,408
Gross profit/loss before income taxes & distributions	$ 457
Allowance for federal income taxes	$ 151
Bonuses	$ 310
Net profit/loss	$ (4)

changing and is only a report of assets, liabilities, and owner's equity on a specific date. It is often said that a balance sheet is a "snapshot in time." It is a very important statement when analyzing the financial health of a company, since it allows calculations or ratios like debit to equity, which can be used to compare a firm with the industry as a whole.

As the simple balance sheet for ACME Surveying & Engineering showed, the balance sheet is primarily concerned with assets, liabilities, and owner's equity. A detailed discussion of what to include in the classification of assets versus liabilities is best left to accountants. In general, firm principals understand that cash and physical things are assets and money owed for immediate or long- term debit are liabilities. This is intuitive and a simple subtraction—assets minus liabilities—seems to produce owner's equity. Owner's equity is what all owners are (or should be) interested in. An increase in owner's equity is an indication that the firm is growing and its value is increasing. It also can give owners reassurance that their effort is worthwhile (increasing OE) or not (decreasing OE). Owner's equity is also equated to the value of a firm and sometimes is known as "book value." This is somewhat of a misnomer since it is only the raw financial value of a firm based on the books. Many other off-balance-sheet items such as the firm's reputation, client list, historic performance, and experience of key staff enter into valuing a firm. A discussion of this is beyond the scope of this chapter. If you are interested in determining the value of your firm, consult an expert who is experienced in valuing surveying and engineering firms.

Assets

I mentioned previously that cash is an obvious asset, so what other items may be considered as assets? The money owed to a firm as *accounts receivable* certainly is an asset, although some adjustments may be needed. Most accounts receivable can be classified as current and aged receivables. Those that are current have a high level of certainty that they will be paid and will quickly be realized as cash. There is little doubt that the total amount of current receivables can be shown on the balance sheet. What about aged receivables? It is common knowledge throughout the financial accounting profession that the older a receivable is, the more likelihood there is that the full amount will not be collected. If a balance sheet's receivables contains a large amount of aged receivables (unpaid over 60–90 days), they should be discounted by the amount that is unlikely to be paid. This is known as an "allowance for bad debits" or "factoring receivables." Other assets include the book value of investments, or "marketable securities," that the firm has, which may include money in stock or money market accounts. They also can quickly be converted to cash. The depreciated value of capital items or "fixed assets," such as vehicles and equipment, also are assets but are not as easily converted to cash.

Depreciation needs some discussion. The reason that a capital purchase such as a company vehicle or office building isn't carried on the balance sheet at its purchase cost is that as soon as use of the asset starts it is no longer as valuable as it was when new. We are all familiar with the saying that an auto is worth $xxxx less the minute you drive it off the dealer's lot. Vehicles, and most other capital equipment, such as survey instruments, wear out or become obsolete and need to be replaced at regular intervals. Depreciation is the accounting process that determines the life of an item and the value subtracted from the original cost of an asset for each year of its accounting life. It assumes that at the end of its accounting life the item has no value. We know that this isn't true, since most vehicles and equipment has some "salvage" or trade-in value. This is one of the adjustments that will be applied by the outside accountant upon reviewing the firm's books.

Another important balance sheet asset is "work-in-process unbilled," commonly known as W-I-P. Why is it so important? There are two reasons. First, when work on a project has been completed, there is real value to the client, whether he has been billed or not. This needs to be shown on the balance sheet as an asset in order to properly reflect an increase in owner's equity even if the client hasn't been invoiced yet. Second, the W-I-P amount shown as an asset also may be subject to adjustment. Often it contains amounts charged to projects that may not be actually be billable. This may be based on time billed to a project that is above that allocated in a lump sum fee or time spent on work outside of the job scope for which there is no agreement. In both cases, there is little likelihood that the amount included in W-I-P will be able to be billed or if billed, will be able to be collected. Since W-I-P and accounts receivable often represent the largest assets of surveying and engineering firms, it is very important that they be adjusted to remove the amount that it is unlikely to collected in order to have a true representation of the asset on the balance sheet.

The value of inventory is generally shown as an asset for retail and manufacturing companies, since there is a considerable amount of money tied up with inventory. This isn't true for professional service firms such as surveying and engineering firm, so generally there is no asset for inventory shown on the balance sheet.

Liabilities

What are liabilities? They are the obligations of the firm that need to be paid. They can be classified as current liabilities and long-term liabilities. They have the effect of reducing owner's equity or book value of a firm. Obvious liabilities include accrued expenses, "accounts payable"' for supplies, equipment, and payroll that hasn't been paid yet. Current liabilities also include the current payments or amounts due on loans or notes that the firm has borrowed to purchase capital equipment. The remaining portion of the note is shown as long-term debit. Another liability maybe an amount that the firm is obligated to pay for employee retirement 401(k) accounts that hasn't been paid yet. There may be other liabilities for which a firm has accrued an amount due but may never actually have to make those payments. These include accrued income taxes and client retainer fees. Contingent liabilities also should be shown on the balance sheet. These may be an amount the firm may be obligated to pay possibly pending a legal action against the firm for which its insurance will not pay.

Stock and Owner's Equity

Why does the liability side of a balance sheet contain stock and owner's equity?

Simply stated, these are the owner's claims against the assets of the firm once all of the liabilities are paid. So what is the difference between the two? When the owner(s) start a firm, they typically put up some cash as their investment to get the firm started. This is known as "common stock" and typically is shown as 100 or 1000 shares. Most small firms are started by one or two individuals with a small investment, so this amount is typically a small number. The exception may be a firm that starts with a large number of partners who put enough money into their initial investment to carry the firm until significant income starts to flow. Often this is enough to cover anticipated expenses for one year. Additional amounts put into the firm by the owner(s) as it grows also may be shown as common stock. Sometimes these amounts are loaned and shown as a liability rather than an investment.

Retained earnings are funds retained from profits to continue the operation of a firm (profits not kept in the firm are paid out as bonuses). The amount retained from profits (increases cash) varies from firm to firm, according to the need to finance ordinary cash flow or growth. Conservative accounting practice includes keeping enough in retained earnings to finance payroll and other expenses for a certain period in order not to have to borrow on a line of credit and incur interest expense. Excess retained earnings become part of owner's equity.

Owner's equity is typically the last item shown on the liability side of the balance sheet. It basically is the amount that is left for the owner(s) once all of the other liability claims are paid. It is easy to see that owner's equity increases if the firm is profitable through an increase in retained earnings and assets or a decrease in liabilities. A significant increase in owner's equity and retained earnings over the long term ensures the owner(s) that their initial investment is growing and generally the venture has been worth the effort.

Through this discussion you now should understand why the balance sheet it just as important as the income statement. As previously mentioned, both need to be understood and used together to obtain a complete financial picture of a firm. A sample balance sheet is shown in Table 7.5.

**Table 7.5 ACME Surveying & Engineering, Inc.
Balance Sheet January 1 to December 31, 20XX
(Accrual - figures rounded to $1000)**

Assets	
Current Assets	
Cash	5
Short-term investments - MM	150
Investments - Mutual Funds	20
Work-in-Process	200
Accounts Receivable	475
Other assets	30
Total current assets	880
Fixed Assets	
Property & equipment	400
Accumulated depreciation	350
Total fixed assets	50
Total Assets	930
Liabilities	
Current Liabilities	
Accounts payable	5
Retainer fees	25
Accrued expenses	100
Deferred Income Taxes	200
Current portion of notes payable	75
Total current liabilities	405
Long-term Liabilities	
Long-term debt less current portion	250
Total Liabilities	655
Owner's Equity	
Capital Stock - 1000 shares	1
	274
Total Owner's Equity	275
Total Liabilities & Owner's Equity	930

CAPITAL SOURCES

Capital is necessary for paying current debits, financing growth, and establishing reserves for unforeseen emergencies. Where does capital come from? Sources include the initial investment of the owner(s) at firm start-up, retained earnings from profits generated, borrowing, and others sources such as outside investors or creating new stock. Let's discuss the advantages and disadvantages of each of these sources.

Managing cash flow through the ups and downs of billing and collection cycles of most small firms is a major task and requires a fiscally conservative and experienced person. Building and maintaining sufficient cash reserves is one of the most important jobs in a firm and shouldn't be delegated to a "bookkeeper." It requires someone who has an excellent grasp of the overall financial needs and condition of the firm. Generally, this is a principal of the firm. Handling and maintaining the cash reserves also includes investing excess cash to obtain the best possible return yet still have it immediately available to satisfy liabilities. Money market accounts are the most common place, outside of the active checking account, to keep excess cash. Additional cash may be able to be invested in bonds or conservative mutual funds, where returns may be higher than money market accounts. Since this is the operating "life blood" of the firm, excess cash should be invested only in conservative, risk-free, investments, where preservation of capital, not return on investment, is the most important consideration.

Borrowing from a bank or other financial institution may be a necessity when cash flow issues arise. This could be the result of having undertaken a very large project for which you can't invoice until certain milestones are met or deliverables are presented. In the meantime, your payroll and other expenses need to be paid, and this need may exceed your normal cash reserves. It is prudent to establish a firm's line of credit before it is needed. The old saying is that bankers are always willing to loan money to those who really don't need it and very skeptical about loaning to those who are in desperate need of money, even if only for a short period of time. When I started in business over 30 years ago, short-term unsecured cash was readily available from our local bank at prime interest rate plus one point. In those days, the local banker knew everyone in town and probably held the mortgage on your house and car loan, so a credit check was as simple as checking to see if you had missed any payments. When necessary, I could obtain such a note just by showing up at the bank and signing a promissory note. In today's financial institutions, this is no longer possible. In fact, short-term credit may not even be available for some firms. Today our line of credit is secured first by the assets of the firm, primarily accounts receivable, and second by the personal assets of the principals. If you don't have assets, you don't get a line of credit.

Whether we use the line of credit or not, each year the bank requires firm and personal financial statements for renewal.

Another source of capital is loans from stockholders, family, or staff. We've used this method in the past to finance large capital purchases since we'd rather pay interest to "members of the family" than a large financial institution. We haven't used this method for short-term borrowing. If you use this method make sure you have the ability to make the loan payments in a timely manner.

Borrowing to bridge short-term cash flow issues or to purchase capital equipment is a normal and necessary part of a firm's operation, and repayment should be planned for in the operating budget. Typically, this is low-risk borrowing with the only real expense being the interest incurred.

Creating capital to finance growth is very different from ordinary borrowing discussed previously. Here risk is high and security may be questionable. Most banks shy away from making loans to finance growth without significant collateral and a detailed business plan. Most firms finance significant growth such as opening a branch office or acquiring another firm by creating additional stock, which they will issue to investors either within or outside of the firm. This type of financing requires the special expertise of investment bankers or venture capitalists, who usually require a significant position in the firm as part of their security. In most cases this type of financing isn't used by small firms.

Getting Paid and Paying Your Bills

The amount of money owed to you is known as "accounts receivable." Everybody who gives credit to clients has them. It isn't bad to have accounts receivable as long as they are kept current and under control. In an ideal world, you would invoice for work completed and would be paid within 30 days, but ideal isn't the world of small surveying and engineering firms. Most firms have large receivables beyond 60 days. In fact, the industry average in 2006 was 67 days. I mentioned previously that the older receivable are, the less likelihood the full amount will be collected. The reason is that, if you assume a client is happy with the services you provided, he or she will pay you within 30 days of receiving your invoice. If the period extends beyond 30 days, it is likely the client is having some sort of financial trouble. This may be a temporary cash flow problem or the client may have failed to secure financing for the project. In either case, stay in close contact with the client and request a regular update regarding when you will be paid. Also be very clear that you are not in the banking business, and you will be charging interest on the outstanding amount. Often this is enough to promote payment. The surveying and engineering profession is well known as a "soft touch," and we generally provide interest-free loans to our clients, sometimes for long periods of time.

Once it becomes clear that you are not going to be paid, it is time for a serious decision, one that should be made by a principal of the firm. You can either take steps to collect the amount owed or you can write it off as a bad debit. If you take steps to collect your money be ready for hostile countermeasures from your client. A notice of collection, or legal action or suit, is likely to cause a counterclaim by the client that your work was negligent, that it wasn't performed in a timely way, or that you didn't provide all of the services contracted for. In any case, be sure you are ready for a long drawn-out battle. Often, this type of collection action takes several years. If you decide to write-off the amount due, it will increase the bad debt expense and reduce profit at best. In the worst case, your firm could become known as one that doesn't take action to collect their receivables and you will attract the deadbeat clients who have no intension of paying right from the beginning.

Paying your bills is the flip side of getting paid. It goes without saying that if you desire a good credit rating and a topnotch business reputation, you should pay your bills promptly. Generally, this is within 30 days of receiving the invoice. Sometimes discounts are given for payment within 5–10 days. Obviously, you should take advantage of these, if your cash flow allows it. At sometime during your firm's business life, cash flow will be slow and you will have difficulty paying your bills. This is the reason for having a previously established line of credit. Such a line is usually activated by a phone call to your local banker, who then deposits the requested amount in your account, thereby allowing you to pay your bills. Most lines of credit charge a fairly high interest rate, and it is expected that the amount borrowed amount will be paid back within 60–90 days. Banks generally require a line of credit to have a zero balance at least once a year. This essentially ensures your banker that he isn't a silent equity partner in your business.

BUDGETS

Why should we budget anyway? The simple answer is to help us with financial decision making but it's not that simple. The budgeting exercise is resisted by almost everyone. Why? Because it's not a financial process; it's a management and planning process. It is not accounting and shouldn't just be left to the accounting staff. It is management, it is strategic planning, it is brainstorming, it is analytical, and it helps us see the future. Budgeting is comparing actual performance, what your firm has done, with what you forecast that you can do.

What are some of the items we should budget for:

- *Income.* Be sure to break it down into real project income versus passthrough items like consultant fees and reimbursable costs.
- *Expenses.* Be sure that you understand the difference between your fixed and variable expenses. Don't spend too much time on your fixed expenses. It is hard (not impossible) to change these.
- *Cash flow.* Do you have a month with higher expenses than others or excessive cash near the end of the year that you don't want to be taxed on? Are high accounts receivable causing the firm to borrow on its line of credit? Will large projects with phase or milestone billing requirements cause cash needs?
- *Utilization (chargeable).* Does everyone have a utilization or chargeable goal? Are they meeting it, and if not how are you dealing with it? Do you know the overall utilization for the firm and what it has to be to be profitable?
- *Profit and bonuses.* Some say these just happen but successful firms plan for them.
- *Capital expenditures.* Do you know what the expenditures for large purchase might be this year? Do you have a capital replacement plan that replicates depreciation? What are the consequences if they are delayed? Will maintenance and repair cost increase?
- *Non-income-producing projects.* Do you budget how much probono work you'll do each year or how much your summer intern, working on a nonbillable project, will cost the firm?
- *Charitable giving.* This one is particularly difficult. If your firm is successful, you become a target and everyone in the firm has their favorite charity. What is a reasonable percent of income to give away each year?

In order to begin the budget process, we need historic information about the past year, and up to five years is even better. We need to understand the circumstances that caused events. Were they internal or external? Did income experience a big increase because of adding key staff that were very billable or did a large project require lots of overtime and produced high utilization? Did we have many small projects that require almost as much administrative and overhead time as large project but don't produce the income or profit?

Conservatism in accounting is required, but it can be damaging in budgeting. It can produce lower expectations and eliminate risk taking that may be necessary for firm growth.

Income Projections

The projected income budget for the year is tied to the projected workload. How do we get our arms around this and how can we project our workload for a year? A crystal ball would be nice, but in lieu of that we have to use a method that combines in-house historic data and the effects of overall economic projections. Our firm does a combination of private, municipal, state, and federal work, so here's how we do it. This method will have to be adjusted for your own firm.

Projecting the private sector workload or income for the coming year is relatively easy. We can compare our historic private sector income with the number of building permits given out by the communities that we work in. If it appears that building permits will be up, then we can project that private sector income will be up. If it looks like building permits will be down, then income in this sector will be down. There's lots of economic data, heard almost everyday, regarding history and trends in housing and commercial construction.

Our municipal workload, including school projects, also is not difficult to forecast, although most of the information doesn't come in until municipal voting takes place in March and April. We work with our municipal clients throughout the fall to help them establish capital projects budgets for the coming year, so we know which projects and how much money will be included in their budgets for total projects, including surveying and engineering fees. Once the voting has taken place and we know which projects and budgets have received favorable votes, we can project our municipal sector income.

The process for state and federal work is similar but sometimes a little vaguer. State budgets, in our state, work on a July 1 to June 30 fiscal year, and usually the legislature is adjusting the budget right down to the closing session. After the budget is approved, it may be difficult to determine which state agencies got how much money and which projects received funding and which didn't. Since we don't do a lot of state work, this doesn't create any serious income budgeting issues for us. The federal fiscal year is October 1 to September 30, which may mean another income adjustment in the third quarter of our year. Our federal work tends to be long duration, 3–5 year contracts, so year to year impact on our budget is minimal. It also is more difficult to find out which projects are funded in the federal budget. The best way is to stay in close contact with the federal agency and the contracting officer that you work with.

Remember to exclude passthrough items, such as consultant fees and reimbursable expenses from your income budget.

Expense Projections

Expense projections may be a little more difficult to forecast than income, but in order to be profitable, we must be able to adjust them as the year progresses. The question is what are variable and what are fixed expenses? I mentioned earlier that it doesn't make sense spending time with fixed expenses. But how do we know which expenses are fixed and which are variable? In some cases, there are degrees of each. Let's start with the obvious ones. In most cases, we have a long-term lease or mortgage payment for our office, which can't be changed without a lot of planning and effort. This is a fixed expense, since it can't be easily adjusted. Other similar expenses include license fees, vehicle registrations, utilities, loan payments on capital purchases, and payroll expenses. Fixed expenses make up a large portion of most firm's budgeted expense. However, some fixed expenses vary widely year to year or can be eliminated for a short period of time if necessary. Variable expenses, on the other hand, by definition can change on a year-to-year basis. The major item for most firms is health insurance. Since we pay the major part of an employee's health insurance premium, it's a big expense in our budget. Each year our business manager shops for the best combination of cost and coverage, but the changes in premium can be 10–30 percent. Other fixed items that can be adjusted include professional society dues, employee education and seminars, charitable contributions, and some forms of insurance such as general and professional liability and vehicle insurance. Another fixed expense is capital purchases. If you need to cut your budget, they can be eliminated for a year and you can make do with the current vehicles, computers and survey equipment. However, keep in mind that maintenance and repair costs may increase because of the need for more frequent repair of capital items that need to be replaced. This is a short-term solution, and firms that continually eliminate capital purchases in order to balance the budget may be headed for trouble in the long run.

The real variable expense is payroll or labor costs, since the number of staff can easily be changed in proportion to the workload. I mentioned previously that payroll expenses may be fixed. This generally is true as long as a staff is in neutral, neither growing or contracting. If you are budgeting for growth and need to add staff, payroll will increase. This should be offset by a proportional increase in income. The portion of payroll expense that is completely variable is bonuses. If the firm is profitable bonuses should be paid as a way to reward the staff and to minimize taxes. If times are tough though eliminating bonuses is a good way to control expenses.

Another expense item that is variable to a certain extent is write-offs for bad debts. These are variable from the point that you may take them at any time and they also can be used as an expense, which helps minimize taxes.

Finally, pro bono work and nonbillable projects are another variable expense that can be minimized or eliminated in declining economic conditions.

Profit Projections

There are two types of profit that we must budget for. We all understand the need to maximize the firm's expenses in order to offset income and therefore minimize taxes, and it is completely legal. The goal of most small firms is to pay no income tax, so we pay out bonuses to staff and make capital purchases of needed items and equipment. This is not the real profit for which we budget. Profit that we budget for is that before discretionary expenses and taxes. In our firm, we budget each project for 20 percent profit and hope for 15 percent. This is slightly above the industry average but it helps us attract and retain superior staff by allowing us to pay good bonuses. Overall firm profit is also affected (and can be adjusted) by controlling overhead expense.

Backlog Projections

Backlog projections are something our firm has been struggling with for a long time. It is necessary that you know your backlog in order to make good budget forecasts but it is an imprecise science. There are two methods we use and since I don't think we've found the perfect one, there may be more.

First, we try to determine backlog on a dollar basis. This is a relatively simple calculation consisting of taking the total value of work under contract and subtracting the value of the work billed and W-I-P. The remaining amount divided by the firm's monthly average billings yields backlog in months. A simple illustration follows:

Total value of work under contract	$ 500,000
Minus total amount billed against contracts	$ 150,000
Minus W-I-P completed but not billed	$ 50,000
Value of remaining work	$ 300,000

If the firm bills an average of $ 100,000 per month
Backlog in months = $ 300,000 / $ 100,000 = 3 months backlog

A similar calculation is done using man-hours remaining on projects as a check on the preceding method. A refinement of these calculations can be to break them down by department or project manager.

These calculations, however, are dependent on good accounting and project management data. The total value of work under contract is only valid if there is a budget for each project and contract amendments, which increase the value, are included. If work for a project is done on a lump sum basis, excess fee for work already completed must be subtracted, since there is no real remaining backlog. In addition, negative amounts for write-offs or work done on projects that are over budget must be changed to zero, since they don't subtract from the backlog either. Work-in-process, unbilled also must be adjusted for write-offs if the calculation is going to be accurate.

In addition to the hard backlog developed for projects under contract, we do a similar projection for the amount of outstanding proposals. We don't use the full value of each proposal unless we know it is a certainty. We assign a probability for success to each proposal amount and multiple the proposal value by that amount. Dividing the probable proposal amount by the average billing per month yields the probable proposal backlog, which is added to the actual contract backlog determined previously. See Figure 7.1 for a sample backlog with adjustments.

The backlog determined is a rough figure, since it does not consider the actual schedule for each project. For example, the 3-month backlog determined in the example may be affected if the contracted amount contains value of construction services that won't be performed for 6 months or more. A more sophisticated backlog calendar that shows the accumulated man-hours

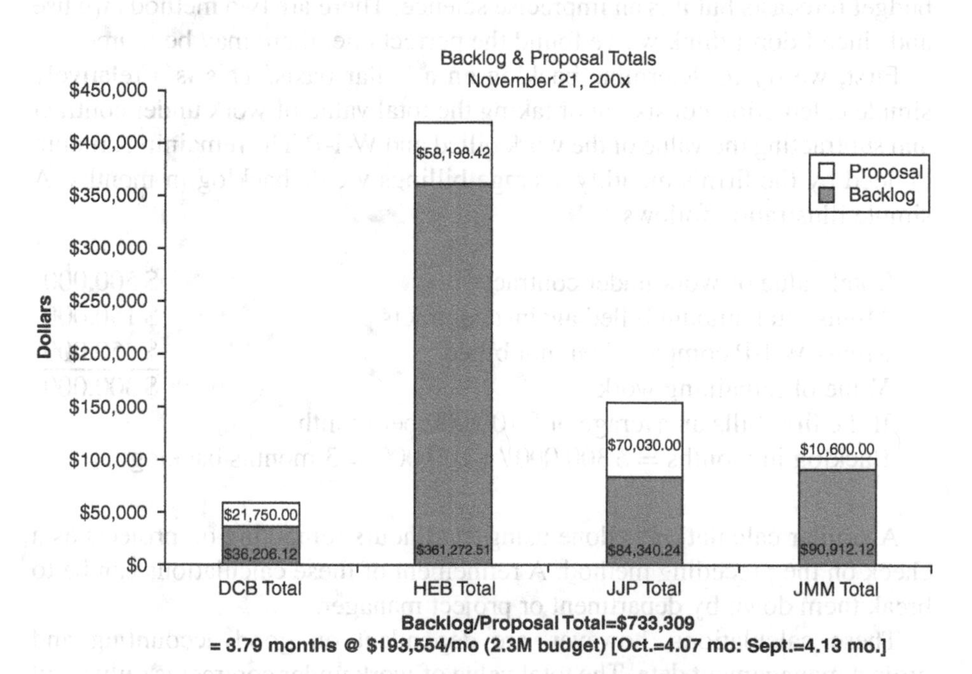

Figure 7.1 Sample backlog adjustments and spreadsheet.

scheduled for all projects is the next step in determining backlog with a greater level of certainty. This backlog schedule will clearly show when resources are under utilized or committed. Because of the complexity of performing this calculation and the amount of data needed (a detailed man-hour schedule for each project), our firm now has had limited success with this calculation. The important thing to remember about backlog calculations is that they are imprecise, and averaging several methods is better than relying on just one.

Measuring Performance–Ratio Analysis

Each firm needs to measure its performance if it is to be successful in the long run. This includes measure actual versus budgeted income and expense for the present year and other parameters such as utilization (chargeable ratio), overhead and several other multipliers, but how does the firm compare to the industry as a whole? Several management and financial companies do surveys of the industry and prepare detailed reports of the industry overall. Key ratios allow us to compare our firm's past performance and also with the industry as a whole. This information is determined by using financial information mostly from the balance sheet and income statement. The common ratios used to track performance of surveying and engineering firms follow. The industry comparative data is from the *2006 & 2007 Operating Statistics Survey*, conducted by Deltek Inc.

a. Debt to Equity Ratio = Total Debt/Owner's Equity (Book Value)

This ratio is a measure of a firm's financial strength. Large firms often have debt-to-equity ratios greater than 1, which indicates that they are highly leveraged. Smaller firms are more conservative and generally seek to have a debt to equity ratio of less than one. Often owners of small firms leave a substantial amount of retainer earnings in owner's equity in order to finance the firm's operations and minimize borrowing. In 2006, the industry average D/E ratio was 0.70.

b. Current Ratio = Current Assets/Current Liabilities

This ratio is an indication of a firm's ability to pay its immediate debt obligations. Current assets are considered to be cash, accounts receivable (possibly deleting questionable doubtful ones over 90 days), and other assets such as investments that can quickly be turned into cash. Current liabilities are accounts payable, accrued payroll, and current portions of loans due. A current ratio greater than 2.0 is considered healthy. In 2006, surveying and engineering firms had an industry average current ratio of 2.88.

c. Chargeable Ratio (Utilization) = Billable Hours Worked/Total Hours Available for Billing, or alternatively, Direct Billable Labor Cost/Total Labor Cost.

No one works the total 2080 (40 hrs/week × 52 weeks) available hours per year. After subtracting vacation, sick leave, and holidays most employees are available to work about 1800 hours per year. Each member of a firm should have a goal for their billable time utilization after subtracting unavailable time. This varies from almost zero for administrative staff to 50 percent for principals and 85 percent for project managers. Most technical staff should be billable 90–100 percent of the time. The overall measure of utilization can be tracked as a trend, which is an indicator of profitability. In our firm, we know that we will be profitable if overall utilization is over 65 percent. The industry for 2006 was 61.2 percent. Utilization also can be calculated on the basis of dollar costs for labor. The cost of billable labor is divided by the total labor cost. The calculation should be similar using either unit.

d. Net Multiplier = Total Income (Minus Passthroughs)/Direct Billable Labor Cost

Net multiplier is an indication of the overall success of a firm (or individual projects) to cover labor and overhead costs and generate profit. If a firm spends $100 for labor and $170 for overhead costs, it needs to multiply labor by 2.7 to break even. If the firm plans to make 15 percent profit on a project labor needs to be multiplied by 3.10 to achieve this. Net multipliers below 3.10 may mean that potential profit wasn't realized ,and below 2.7 means that costs were not even covered. In 2006, the industry average net multiplier was 3.0.

e. Chargeable Ratio × Net Multiplier is a new measure that has recently been adopted by some firms. It can be used as a single indicator of effective utilization and overall profitability shown by the net multiplier. It also can be used as an indictor of a high level of write-offs. If the chargeable ratio is high yet net multiplier is low, it is an indication that although employees are charging their time to projects, all of the time cannot be billed.

f. Overhead Multiplier = General and Administrative Costs/Direct Billable Labor Cost

Overhead costs have been increasing over the years. The need for more technological equipment like computers, cell phones, high-speed Internet access, and electronic surveying equipment has contributed to the increase. Ongoing costs like upgrading and relicensing software and increases in professional liability and health insurance and also increasing fuel costs increase overhead. General and administrative costs used in the equation are always minus discretionary distributions like bonuses but include payroll burden such as employers share of FICA and state unemployment insurance cost. The increase in overhead can present problems with clients (mostly state and federal agencies), who limit overhead to a certain maximum. In 2006, the industry average was 1.55. Sometimes overhead is stated as a percent of labor cost. In this case, it would be 155 percent.

g. Profit (before Discretionary Distributions) = Net Income/Total Fee Income (Minus Passthrough Items)

It is important to subtract passthrough items like consultants and reimbursable costs from total fee income, since they can skew profit downward. In 2006, the industry average was 13.8 percent.

h. Net Revenue per Technical Staff = Total Fee Income/Number of Professional Staff

This number is an indicator of trends. An increase indicates increasing profitability, while a decrease indicates an increase in write-offs and rework. The amount also varies geographically, since lower billing rates produce less total fee income and, therefore, lower net revenue per technical staff. In 2006, the industry average was $136,283. A variation of this is net revenue per total staff, where total fee income is divided by the total number of people on staff.

i. Average Collection Period in Days = Total Accounts Receivable/Average Billing per Day

This is the average number of days it takes to collect money owed to the firm. It is directly related to how long a firm is willing to extend credit to its clients. Generally, firms will extend credit for 30–60 days and then charge interest on outstanding balances. The Deltek survey indicated that there is a declining trend in firms charging interest on overdue accounts. This effectively provides clients with interest-free loans. In 2006, the average collection period was 67 days.

The preceding are the most common ratios used throughout the surveying and engineering industry. There are additional ones that provide good

Table 7.6 ACME Surveying & Engineering, Inc. Ratio Analysis January 1 to December 31, 20XX

	ACME	Industry Ave.
Debit to Equity Ratio	0.70	0.70
Current Ratio	2.17	2.88
Chargeable Ratio (Utilization)	65%	61.2%
Net Multiplier (excluding passthroughs)	3.11	3.00
Overhead Multiplier	1.35	1.55
Net Profit (before discretionary distributions)	25%	13.80%
Net Revenue per Technical Staff	$103,611	$100,000
Ave. Collection Period (days)	60	65

information but are used less frequently and, therefore, it may be harder to compare your firm's performance with the industry. Most managers and executives do not utilize all of the ratios previously outlined. They find several key ratios, which in their mind are trustworthy indicators of the firm's performance. The following table shows how ACME Surveying & Engineering Compares with the industry.

Ratios also can be used to perform "what if" projections. The following examples show the results of increasing the chargeable ratio 3 percent and increasing the net multiplier from 2.95 to 3.0. In both cases, net profit increases substantially.

Breakeven Analysis

A breakeven analysis is a method of calculating how much income is necessary to cover all of a firm's costs. It is another tool that helps management in financial decision making. It can help determine billable rates and in budgeting all costs of operation, including both fixed and variable expenses. A review definition of the two types of expense is as follows:

Fixed expenses. Are those cost that do not change with changes in work volume, productivity, or income. They stay the same no matter what the economy or the firm workload does. They are rent, utilities, depreciation, administrative staff costs, insurance, and other general administration expenses. Most overhead expenses are fixed, and there is little to be gained by trying to cut fixed costs.

Table 7.7 Ratios & "What If" Calculations
ACME Surveying & Engineering, Inc.
Ration Analysis January 1 to December 31, 20XX

Number of employees	20
Net revenue per employee	$100,000
Total net revenue	$2,000,000
Net Multiplier	2.95
Calculated direct labor cost	$677,966
Chargeable ratio (Utilization)	62%
Calculated salaries & wages	$1,093,494
Total salaries & wages as % of net revenue	55%
Calculated overhead cost	$1,072,034
Overhead %	158%
Net profit $	$250,000
Net profit before distributions	
Direct labor	1.00
Overhead	1.58
Profit	0.37
Total net multiplier	2.95
What if chargeable ratio is increased from 62% to 65%	
Number of employees	20
Net revenue per employee	$104,839
Total net revenue	$2,096,774
Net Multiplier	2.95
Calculated direct labor cost	$710,771
Chargeable ratio (Utilization)	65%
Calculated salaries & wages	$1,093,494
Total salaries & wages as % of net revenue	52%
Actual overhead cost	$1,039,229
Overhead %	146%
Net profit $	$346,774
Net profit before distributions	
Direct labor	1.00
Overhead	1.46
Profit	0.49
Total net multiplier	2.95
What if net multiplier increased from 2.95 to 3.0	
Number of employees	20
Net revenue per employee	$101,695
Total net revenue	$2,033,898
Net Multiplier	3.00
Direct labor cost	$677,966
Chargeable ratio (Utilization)	62%
Calculated salaries & wages	$1,093,494
Total salaries & wages as % of net revenue	55%
Actual overhead cost	$1,072,034

Table 7.7 (*Continued*)

Overhead %	158%
Net profit $	$283,898
Net profit before distributions	
Direct labor	1.00
Overhead	1.58
Profit	0.37
Total net multiplier	2.95
Direct labor	1.00
Overhead	1.58
Profit	0.42
Total net multiplier	3.00

Variable expenses. Are those costs that change with the volume of work. By this definition, direct (billable) labor is a variable expense, as well as those benefit costs associated with direct labor. Sometimes management, because of a reluctance to reduce labor cost during an economic decline, treats it more like a fixed cost. That is, direct labor becomes indirect (nonbillable) labor during times when staff is working on nonbillable projects just to stay busy. Obviously, this increases overhead. Other variable expenses are travel, and sometimes outside legal and accounting costs. To a certain extent some items such as supplies, vehicles, and equipment repair costs can be delayed and, therefore, become variable expenses.

It also should be understood that there is not a direct relationship between expenses and increasing fee income. In fact, fee income, which is a function of productivity, may be increasing or decreasing while expenses stay constant. For example, if a firm's physical plant size stays constant but additional office space or work stations are added by changing the floor plan, more profit should be generated, since overhead should remain constant. See Figure 7.2 for an example of breakeven analysis.

This chapter has introduced you to the basics of financial accounting, which should be helpful in operating your firm more profitably. It is not intended to make surveyors and engineers into bookkeepers and accountants. It is intended to help you understand the financial operations of your firm and decipher the two most important financial statements, the income statement and the balance sheet. If you currently receive statements quarterly or annually from your outside accountant, you should consider setting up your in-house accounting system so that you can produce them monthly yourself. There are many software packages available that integrate the time accounting system into the financial management system.

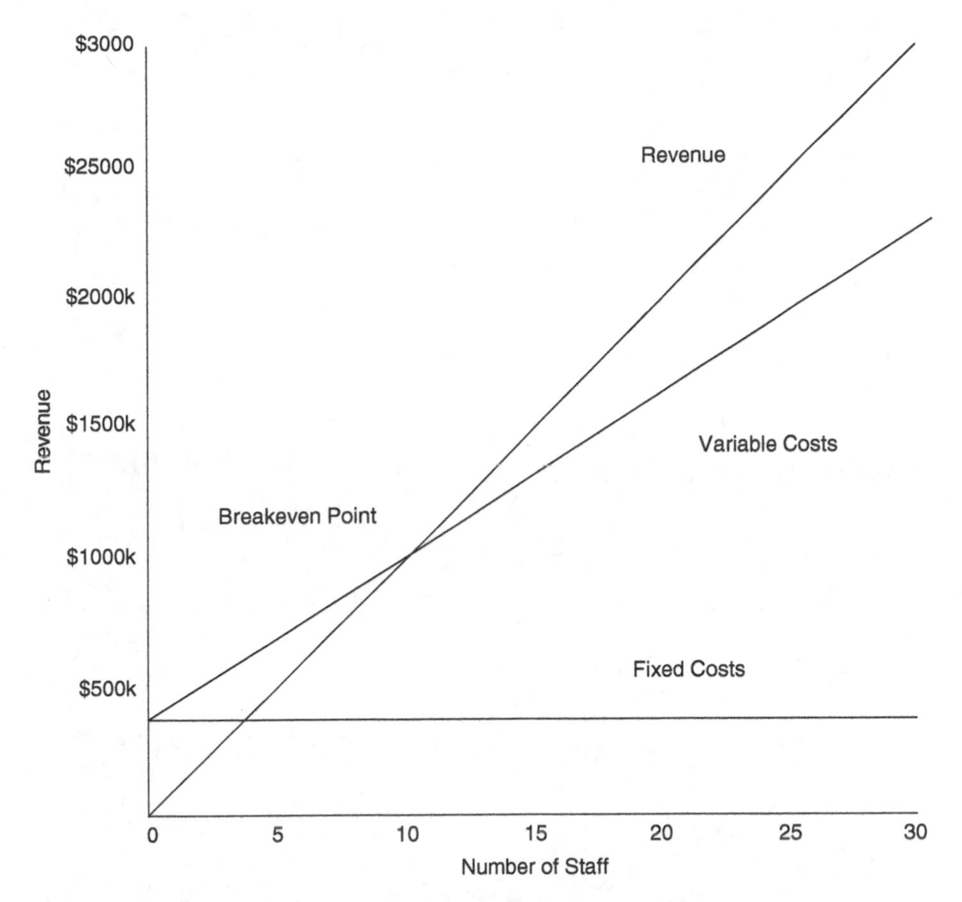

Fixed Costs = Costs which do not vary due to number of employees and revenue.
Variable Costs = Costs which vary with revenue volume. (direct labor etc.)
Breakeven Point = 10 employees, $1,000,000

Figure 7.2 Breakeven example.

Following the introduction to accounting basics, I discussed how to budget, calculate key ratios, and compare your firm to the industry as a whole. This process, known as managerial accounting or financial analysis is intended help managers understand the financial issues that can affect a firm's day-to-day operations. Once you have conducted an analysis of "the numbers," you can compare your firm's performance with industry standards or averages, which are provided by several organizations that track performance of the surveying and engineering industry. If your firm isn't performing up to the rest of the industry, you now have the tools to perform "what if" calculations to explore the results of improving some of your ratios.

I hope you can relate to the financial information for ACME Surveying & Engineering, a firm which I'm sure is similar to many of yours.

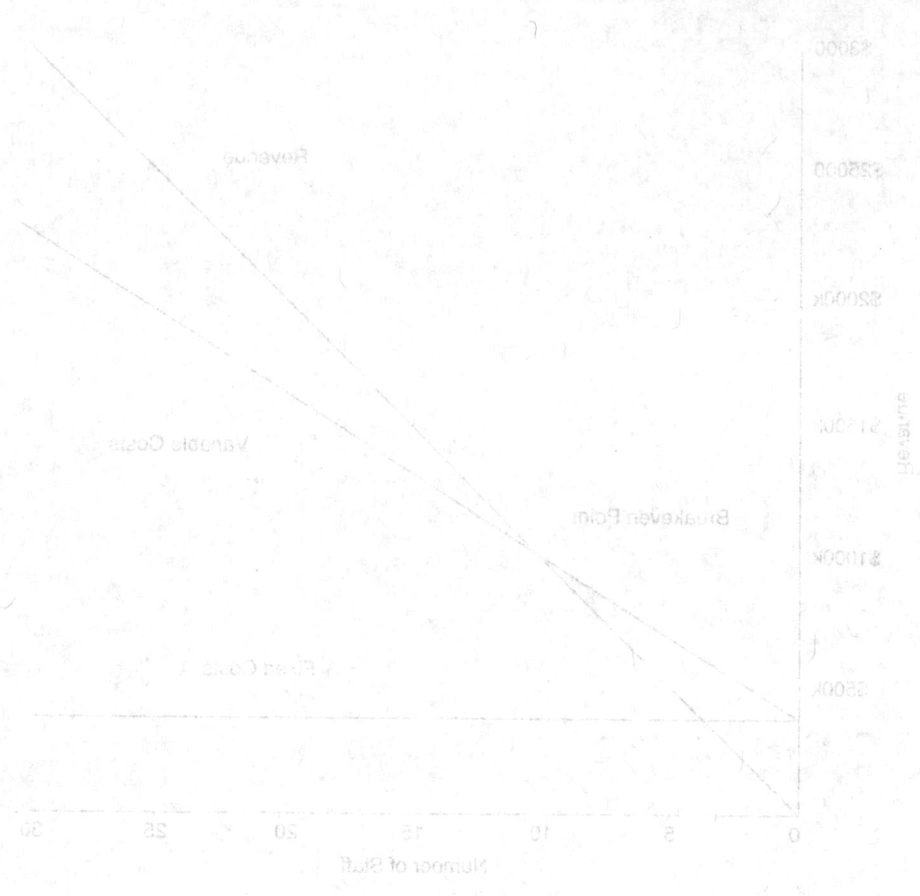

8

LEGAL ISSUES

"You're not a professional until you've been sued."

—Anonymous

INTRODUCTION

Before we start, I need to issue the standard disclaimer. If you've read anything else in this book you know I'm an engineer, not a lawyer, so anything you read in this chapter shouldn't be construed as legal advice. My statements and examples are general in nature based on many years of interfacing with attorneys and the legal system. You should seek legal advice from your own council for your specific legal issue.

Some say, "you're not a professional until you've been sued." Well I guess I'm a professional then. This not to say that our firm has or has not made mistakes in our 34 years of being in business just that we've been the target of a few law suits. Whether you've made a mistake or not sometimes have little to do with being sued. That's why we have professional liability insurance. But there's a lot more that engineers and surveyors need to know, so let's get into it.

DEFINITIONS

To get started, there are some important legal definitions that engineers and surveyors should be familiar with:

Plaintiff. This is the person or other entity that brings a legal action against another.

Defendant. The person or other entity that defends the action brought against them. Generally, the title of a case or action has the plaintiff's name first and the defendant second; for example, *John Brown v. ACME Engineering and Surveying*.

Appellant. A person or party who appeals a trial court decision. The party seeking to uphold the trial court decision is known as the appellee or respondent.

Breach. This is a violation of a duty to perform an act or follow through on an obligation. It is a very important aspect of the law for engineers and surveyors, since many actions against them have a claim for breach of contract.

Claim. This is usually a request for additional compensation based on a contractor's claim that work performed was not included in the contract. Claims are often filed when the contractor is getting into financial trouble on a project or when his relationship with the owner becomes confrontational. Claims are often confused with requests for change orders, but they are not the same.

Change Order. These are requests made in the normal course of a project when the owner or contractor requests that some condition of the contract be changed. It may be that the owner has decided to change the type of windows and doors for his home that is under construction or a contractor may suggest different pumps or electrical equipment in a pump station as the project progresses. Change orders can be for price increases or decreases, or a no cost change may be requested to seek an extension of the contract completion date.

Common Law. This is legal doctrine, which has its origin in court decisions rather than being found in statutes or regulations. This is the reason attorneys often site previous cases in their documents.

Contract. An agreement between parties to perform a specified action, task, or project. A contract can be specific or implied, and it can be written or verbal. In my mind, fewer problems arise when an agreement is reduced to a written document. Even then there may be questions of interpretation. If they can't be resolved between the parties, this is where arbitrators, mediators, or lawyers come into the picture.

Contract of adhesion. This is when one party submits a "take it or leave it" contract to the other party. Adhesion contracts may be considered onerous on the other party, since they may have a weaker bargaining position and may not be capable of understanding all of the terms and conditions. Engineers and surveyors can be found on both sides of these contracts.

Duty. This is a requirement to perform. It often is another claim in legal actions.

Damages. This is compensation usually awarded by the court or part of a settlement agreement to the party who has been wronged. This is where engineers and surveyors often become parties to damage awards whether they are liable or not.

Deep Pocket Theory. You may not find this in a law dictionary. When a court determines that a party has been wronged and needs to be compensated or "made whole." The court generally looks to the defendants in an action to compensate the damaged party according to their ability to pay, not whether they have actually committed a negligent act. This is often the case when a personal injury accident causes a person to be paralyzed, unemployable, and/or have significant medical expenses for the rest of their lives.

Fraud. An intentional and deceitful act or practice depriving another person or party of his/her rights or causing injury in some respect.

Liability. This is being bound or obligated according to the law. It includes general and professional liability.

Negligence. This is a breach of duty to exercise requisite care and expertise. Generally, negligence is considered the exercise of care or expertise below an expected standard or norm. The standard or norm may be hard to prove.

Standard of Practice. The normal practice of reasonable professionals in a similar situation. This often is a hard concept for clients and the public to understand. Professionals, whether they are doctors, lawyers, accountants, architects, or engineers and surveyors are not held to a standard of perfection or even a level of expertise above that which is the standard of the industry. Sometimes clients expect otherwise.

Statute of Limitation. This establishes a time beyond which the parties are no longer liable for damages arising out of a completed project. This is a very tricky situation, since in some cases the statute of limitations begins when the problem is found, which can be many years after the completion of construction. Check your state statute and consult your attorney.

Summary Judgment. A judgment entered by a court when no substantial dispute of fact exists; consequently, there is no need for a trial.

Tort. Any and all wrongful acts done by (claimed) one person to the detriment of another but only those for which the victim may demand legal redress. A tort may be committed intentionally or unintentionally with or without force. It is a private injury rather than a crime. Torts can cause major legal issues for engineers and surveyors, since any third party may bring a claim of an injury. Professional liability insurance companies pay out many thousands of dollars to settle tort claims where there is no liability. In my opinion, this is one of the major reasons for rapid increases in professional liability insurance costs.

These are only a few definitions of importance to engineers and surveyors. Consult a law dictionary such as *Black's Law Dictionary* for others.

THE LEGAL SYSTEM

Sources of Laws

The Constitution. Both federal and state constitutions form the basis of many of our laws and rights as citizens. It defines the separation of powers between the executive, legislative, and judicial branches and the protection of citizens from abuse of power by the government. While most legal issues affecting engineers and surveyors don't rise to a constitutional level, it serves as the basis for many state and federal laws and regulations under which we practice. The federal system includes laws, which are set up and govern federal agencies, which we deal with regularly such as the Army Corp, EPA, EDA, FAA, OSHA, FHWA, USRDA, and the TSA. Those rights not specifically granted to the federal government are the rights of the individual state.

Legislation. Can play an important part in the day-to-day operations of engineering and surveying firms and the design and construction process as a whole. State laws create the licensing process, which governs our professional practice and in some cases professional ethics. State statutes generally give state agencies the power to promulgate administrative rules that directly affect our clients and their projects. In most states, legislation also defines the statute of limitations, the right of communities to develop planning, zoning, and subdivision regulations and to adopt building codes. Legislation is generally considered the most democratic process, since there is an opportunity to elect our public officials and often individuals participate in the public hearing process as laws and regulations are being considered. Legislation can be enacted relatively quickly, thus responding to the social and economic needs of citizens.

Common law. Another source of law is common law. The courts make common law each day through their decisions. If they are appealed and withstand the test, they set precedence and are often referenced by attorneys in establishing a basis for their arguments. The process by which the appellate court judges the correctness of the trial court decisions has a big impact on common law being used as the basis for design and construction cases.

Administrative agencies. These agencies (both state and federal) create regulations based on powers given them in a specific law. Agencies and experts in the particular field better regulate certain activities and industries. They develop better regulations than the legislature because of their expertise and also because of the legislature's unwillingness or inability to get involved in the details. Regulations promulgated by agencies and administrative experts have generated a lot of controversy. They also play an important role in the engineering and surveying industry through the environmental permitting process.

Other sources. The executive branch (the president at the federal level and governor at the state level) can create laws by executive order. Some include orders affecting wages, dispute resolutions, minority and women set asides, pardons, and decrees. Finally, the contracting parties can make law. They have broad autonomy to determine the conditions of their contract, but they cannot contract for something, which is illegal. State law determines who can contract. This generally includes those who have reached the age of majority and legal entities such as corporations, partnerships, and trusts.

The Judicial System

What are some of the courts where engineers and surveyors may end up?

Small claims courts. These are local district courts where minor legal issues are dealt with. Often the plaintiff and defendant represent them selves. The proceedings, therefore, are more informal than in other courts and the cases heard of less significance. Engineers and surveyors use small claims court to collect debts owed. There generally is a limit on the dollar amount of a claim. In our state it is $5000.

State trial courts. A state trial court is the county superior court, in many states, where most engineering, surveying, design, and construction cases are litigated. Trial may be before a judge known as a special master or it may be a jury trial. There are advantages and disadvantages to both. A special master often is a judge with experience and interest in engineering- and surveying-related cases. Rulings tend to be in strict accordance with the law and most are not challenged on appeal. Jury trials tend to be longer and more costly, and many believe that jury decisions are based more on emotion than on the rule of law. Preparation for trial may be extensive and include pleadings, interrogatories, discovery, depositions, motions, hearings, and selection of expert witnesses. Jury selection adds additional time and expense.

Appellate courts. This is where people go to appeal a decision of a trial court. In my state, appeals go directly to the state supreme court, but it is not an automatic process. The Supreme Court first decides if there was a probable error of law at the lower court level. They can remand a case back to the lower court, or they can decide to hear it.

Federal courts. These courts decide federal questions, including disputes or question concerning the Constitution. The federal court will hear disputes between citizens of different states which is the only likely cases involving the design and construction industry. Occasionally, engineers and surveyors will be asked to testify in Federal Bankruptcy Court. The first federal trial court is the Federal District Court. Appeals are made to the circuit court of appeals. The final appeal is made to the U.S. Supreme Court, which is unlikely in a design or construction cases.

Judges may be appointed or elected. In my state, the governor appoints all judges. The president appoints federal court judges. There are good arguments for and against both systems.

Alternatives to Court

> "Who's in control of cases? In litigation, the lawyers think they are in control, judges think they are in control and in some cases clients think they are in control. In reality, no one is in control."
>
> —Barbara Ashley Phillips

The average civil case takes three to five years to prepare and litigate, and 90 percent of the civil cases are settled before going to trial. Since court trials are time consuming and costly many engineering and surveying related cases choose an alternative dispute resolution method. These have the advantage of

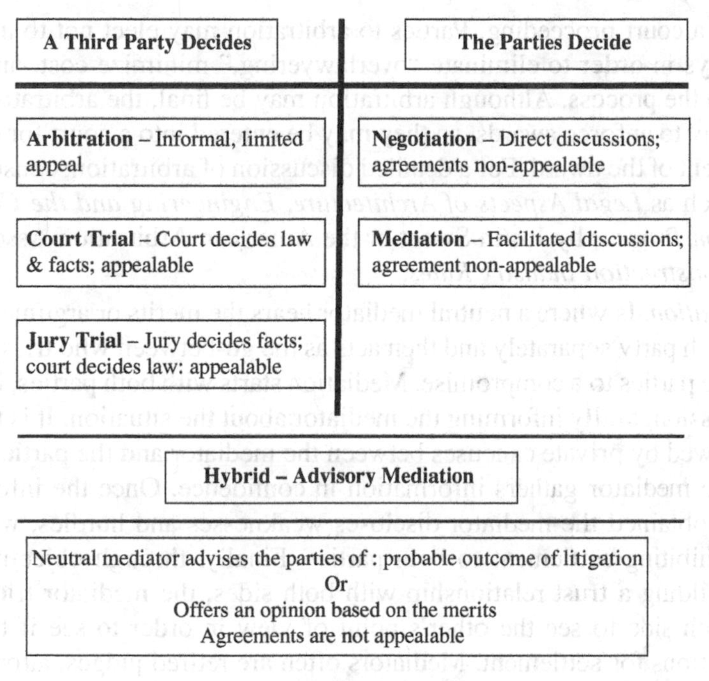

A Third Party Decides
The Parties Decide
Arbitration – Informal, limited appeal
Negotiation – Direct discussions; agreements non-appealable
Court Trial – Court decides law & facts; appealable
Mediation – Facilitated discussions; agreement non-appealable
Jury Trial – Jury decides facts; court decides law: appealable
Hybrid – Advisory Mediation
Neutral mediator advises the parties of : probable outcome of litigation
Or
Offers an opinion based on the merits
Agreements are not appealable

Figure 8.1 Methods of dispute resolution.

being quicker and less costly, and generally are heard by experts in the field. The various types are shown in Figure 8.1 and described here:

Arbitration. In this approach, a dispute is submitted to one or more mutually agreed upon third-party arbitrators, who will hear the case and make a determination. The parties should question the arbitrators in order to be assured that there are no matters that will affect their impartiality. Arbitration is often specified in contract agreements as the first means of dispute resolution. Normally, it is manifested in the contract document signed by each party. Arbitration arose in the 1960s and 1970s as an attempt to move the backlog of civil cases out of the trial courts to make room for criminal cases. It can be binding or not binding, but in practicality arbitration is final. Arbitration should be used to hear cases of factual disputes involving technical standards or issues. This is why it is well suited to engineering and construction disputes. The arbitration clause may not be enforceable if it is viewed by the court as providing more of a benefit to the party preparing the contract than to the other, weaker party. In engineering and surveying cases, it is generally conducted in accordance with the rules of the American Arbitration Association. Some attorneys don't like arbitration, since the rules of evidence, discovery, and testimony are not as well defined, as in a court proceeding. Parties to arbitration may elect not to use attorneys in order to eliminate "overlawyering," minimize cost, and speed up the process. Although arbitration may be final, the arbitrator has no way to enforce awards, so they may be entered into a court for enforcement of the award. For a detailed discussion of arbitration, consult a text such as *Legal Aspects of Architecture, Engineering and the Construction Process* by Justin Sweet or the American Arbitration Association, *Construction Industry Rules.*

Mediation. Is where a neutral mediator hears the merits or arguments from each party separately and then acts as the go-between who tries to bring the parties to a compromise. Mediation starts with both parties, in a joint session, orally informing the mediator about the situation. It is then followed by private caucuses between the mediator and the parties, where the mediator gathers information in confidence. Once the information is obtained the mediator discloses weaknesses and hurdles, which are inhibiting settlement to both parties. Finally, through objectivity and building a trust relationship with both sides, the mediator tries to get each side to see the other's point of view in order to see if there are options for settlement. Mediators often are retired judges, attorneys, or other professionals who are experienced in the surveying and engineering or construction industry. Their goal is to bring both parties together as quickly as possible or to determine that a compromise will not be possible. Mediation works well in the design and construction industry and may be replacing arbitration as the preferred alternate means of dispute resolution. In some locations, the court system has mandated that mediation be attempted before trial in order to encourage settlement and reduce caseload on the court. In mediation, the parties are in control, not the system.

Settlement. Often design and construction claims involve attempts at compromise so that court or alternative means of dispute resolution can be avoided. Settlement attempts can be very time consuming if one of the parties is more interested in delays than in resolving a dispute.

Typically, alternative means of dispute resolution are faster and cost less than regular court proceedings. For this simple reason, it is often preferred over court. I recommend that you always consult your attorney and professional liability insurance company before electing any method of dispute resolution.

Parties to Litigation The primary parties to litigation include the *plaintiff* and the *defendant* previously defined. The plaintiff begins the action but the defendant often files counter claims in an attempt to neutralize the initial claims and provide the first basis for attempting settlement. There can be multiple plaintiffs and defendants in a case. For example, a subcontractor sues the general contractor for unsafe conditions when a worker is injured. The general contractor may, in turn, sue the engineer claiming defaults in the drawings or specifications caused the injury to the subcontractor's worker. The injured worker likely will sue all of the parties, including the contractor, subcontractor, owner, and engineer. These are known as cross claims, and the engineer would become a third-party defendant. Courts often seek to resolve certain disputes by combining related claims in one lawsuit.

Pleadings Pleadings inform each party of the other's claims or contentions. They often are delivered in a summons. If an attorney is unsure of the facts he may plead a number of different legal theories for protection of his client, to cause a delay or to try to initiate settlement. Exaggerated statements and overstatements may be used in the pleadings. If the defendant does not respond to the pleadings, the judge may make a default judgment for the plaintiff.

Pretrial Activities Pretrial activities include a variety of items, which may or may not be used by either the plaintiff or defendant. They include; interrogatories, discovery, declaration of experts, depositions, pretrial conferences, motions, and jury selection.

Interrogatories are questions asked by one side, either the plaintiff or defendant, to attempt to bring out the facts in a case. Questions are often vague, and sometimes it seems like the other side is fishing. They must be responded to within the specified time or the other side risks loosing by default.

Discovery is the process of requesting information in order to clarify the facts of a case and also to obtain information needed for the trial or for an expert witness to conduct his investigation. Almost everything in an engineer or surveyors file, except certain confidential documents, is discoverable and, therefore, available to the opposite side. This is an excellent reason to keep your project files orderly. It can save many hours during discovery.

Declaration of experts or expert witnesses is the process in which each side notifies the other of who their experts are. Sometimes one side in a case will engage an expert, and the other side will not. An expert's report is almost always discoverable.

Depositions are pretrial questioning sessions by one side in an attempt to discover what a witness will testify to or how they will answer certain questions during trial. Expert witnesses are often deposed, and their answers to key questions may help a case be settled before it ever comes to trial. Depositions are always recorded by a stenographer. A copy of the transcript will be sent to you so that you can review your testimony and correct any wrong recording of your responses to questions. Once you receive your copy of the deposition transcript be sure to review it and return it within the specified time period, usually 30 days; otherwise, it will be assumed that there are no corrections. Be sure to review your deposition before trial, and always answer questions asked at trial exactly as you did during your deposition. It is a favorite trick of attorneys to try to get you to answer a question differently during trial from the way you answered the same question during your deposition. He will instantly point this out in an attempt to ruin your credibility.

Pretrial conferences are often held to attempt to limit the issues, which will be argued, and keep the trial within defined boundaries. These conferences are sometimes known as structuring conferences.

Motions are made before the trial and sometimes during a trial. They usually are an attempt to seek a judge's opinion on a rule of law. Sometimes they are a motion for summary judgment or mistrial.

Jury selection is a pretrial event where the jury for a trial is selected from a pool of potential jurors. Many potential jurors may be questioned regarding their knowledge of the case and relationship to any of the parties, or to determine any other fact that may affect their ability to make an unbiased assessment of the fact and judgment in the case. Most cases in the engineering, surveying, or design and construction industry do not use jury trials.

Request for a Master or Jury Trial In most cases, the right to be heard before a jury is a constitutional right, which can't be denied. However, jury trials usually are more time-consuming, more costly and not well suited to the engineering and surveying cases. It also is generally agreed that jurors arrive at their decision based more on emotion and theatrics of presentation of the case than on the specifics of the rule of law. Jury trials in the design and construction industry are generally considered a "crap shoot." They generally are used when a case involves personal injury.

In most cases in our industry, the parties will elect to have the case heard by a judge who is experienced in the engineering, surveying, and construction industry known as a special master or just a master. This generally ensures that the decision will be based on the merits and decided in strict accordance with the law.

Contract law

Basics of Contracts The freedom to contract assumes that the parties are relatively equal, of legal age and mentally aware. In addition, an individual who contracts on behalf of a corporation, partnership or trust must be an officer, partner, or assigned legal power, usually by the board of directors, to act on behalf of the organization. Many government agencies require specific proof in the form of a "power of attorney" or corporate resolution, that the person signing the contract has the authorization to do so. This is attached to the contract agreement. It also assumes reasonable expectations that parties will perform as promised. American law gives parties broad autonomy to choose the content of their contract as long as it is legal. Most engineering and surveying contracts are economic exchanges. In other words, you provide an engineering or surveying service for a fee paid by the client.

Beware of mass-produced or "take it or leave it" contracts (aka contract of adhesion) provided by the client. They may be a purchase order where the terms and conditions have little or no relevance to engineering or surveying services. An example is a condition requiring delivery of and item to include transportation, "FOB the loading dock at ACE manufacturing Company" or "10% discount if payment is made within 30 days." Many of these conditions are slanted to the benefit of the client and often are not applicable to the engineering and surveying industry. Obviously, the best position to be in is to be able to walk away from such agreements. Sometimes you may not be able to. If this is the case, be sure to weigh the potential benefits versus the costs. Read and understand the terms and conditions and consult your attorney and professional liability insurer.

Many years ago we had such a situation with a "take it or leave it" contract. We were selected for a large contract by a telecommunication company. They furnished us with a contract, which contained terms, and conditions, which our insurer reviewed and said, were not appropriate for a professional services agreement. The client insisted that they were fair and developed by their legal department. Our response was that we were not going to sign the agreement until the insurance company attorneys worked out the terms and conditions with the company's legal department. At the project level, we agreed to proceed based on our firm providing a detailed scope of work for each individual work order and attaching it to our standard agreement with our terms and conditions. It ultimately may have been determined that we were working without a contract, since they never signed and returned our agreements, but we proceeded on that basis for several years. To the best of my knowledge, the attorneys never worked out the terms and conditions.

The engineering industry, through the Engineers Joint Contract Document Council or EJCDC, has developed a set of contract agreements, general conditions and other contract provisions, which were developed over the years by consensus and are widely accepted as fair and enforceable.

LETTER AGREEMENT

Date:	January 2, 200X	Job No:	200X- 001
To:	John Director Fineville Public Works Department Pothole Drive Tretorn, AK 10009	Phone: Fax:	(603) 555-5555 (603) 555-5556
From:	ACME Engineers & Surveyors	PM: Real Reliable, P.E.	
Re:	Notwell Bridge over Swollen Creek		

Dear Mr. Director,

We propose to render professional engineering and surveying services in connection with repair, rehabilitation, or replacement of the Notwell Bridge, hereinafter called the "Project." You are expected to furnish us with full information as to your requirements, including any special or extraordinary consideration for the Project or special services as needed, and to make available all pertinent existing information.

Our scope of services will consist of: *An engineering study to determine the extent of damage to the bridge, evaluate alternatives for repair, complete rehabilitation or replacement of the bridge. This includes survey base mapping, geotechnical investigation, wetlands delineation, archeological and historic investigation, exploration of up to three alternates, including engineer's opinion of probable cost estimates for each.*

Deliverables: Three copies of the completed report.
Meetings: Two review meetings with Mr. Director.

See the General Provisions (Terms and Conditions) on the back of this page for a more detailed description of our and your obligations and responsibilities.

You will pay us for our services on an hourly basis in accordance with fee schedule in effect at the time services are rendered, estimated at $ 10,000 to $ 12,000, plus reimbursable expenses.

A retainer fee in the amount of $ 500 would be appreciated at the beginning of our work.

We would expect to start our services promptly after receipt of your acceptance of this proposal and to complete our work within 2–4 months. If there are protracted delays for reasons beyond our control, we would expect to renegotiate with you the basis for our compensation in order to take into consideration changes in price indices and pay scale applicable to the period when our services are in fact being rendered. This proposal is void after 30 days.

This proposal, the General Provisions, and the fee schedule represent the entire understanding between us in respect to this Project and may only be modified in writing and signed by both of us. If you agree with these arrangements, we would appreciate your signing one copy of this letter in the space provided below and returning it to us along with the requested retainer.

Accepted this _____ day of _____ 200X	Very truly yours, ACME Engineers & Surveyors, Inc.
By: ________________________	By: ________________________
Printed Name/Title: ______________	Title: ________________________

Types of Contracts Contracts can be expressed or implied. Expressed contracts manifest the agreement and can be either written or oral. Implied contracts manifest acts rather than words. Obviously, a written contract, which specifies in detail, the agreement between the parties is the best form of contract.

Some contract terms to be aware of include:

- At all times, all information, all permits, all meetings, and the like
- Highly trained or expert professionals
- Is exceedingly well qualified
- High-quality reports
- Only the best
- We guarantee successful completion of the project
- Only experience and qualified staff will be assigned
- Insure, ensure, assure
- Certify, warranty, guarantee

Another contract item to keep in mind is the schedule. Sometimes an aggressive schedule is presented based on the client's desire to obtain results as quickly as possible, but the client doesn't sign the agreement in a timely manner, and the schedule becomes unrealistic right away. You should seek a contract amendment right away, since not performing in accordance with the schedule is a technical breach.

BASICS OF TORT LAW

A Private Act

A tort is a private act wrongfully done by one person to another for which only the victim may claim legal redress. It may be intentional or unintentional, and it may be an injury to a person, property, or reputation. A tort is distinguished from a crime in that it is a private affair or injury. A crime is against the public for which retribution must be sought by the appropriate government authority.

Why are torts important for engineers and surveyors? This is where third-party actions arise from a job site injury or some other situation where a person who has been harmed who has no contract with the party being sued. There are more third-party tort claims brought against engineers and surveyors than contract claims—for example, slip and fall cases or deep pocket awards.

Potential third parties in torts are:

- Contractors/subcontractors
- Construction workers
- Material and equipment suppliers
- Sureties and insurers
- Owners
- Tenants
- Employees
- Visitors
- The public in general, who feel they have been harmed

Torts Compared to Contracts (Breach versus Tort Claim)

A breach requires proof that the defendant has not done what was promised. A tort claim requires proof that the defendant didn't perform in accordance with the law. A failure of an engineer or surveyor to perform according to their agreement with a client may be breach. The duty owed in a contract claim is to the other party in the contract. In a tort claim, the tortfeasor (defendant), owes a duty to the claimant based on the feasibility that his acts or failure to act exposed the claimant to unreasonable risk. Claims against engineers and surveyors are almost always economic and created by contract.

The law uses torts to protect against harm to persons or property. It is reluctant to use torts to protect against an economic loss. Often the claimant can elect to use either contract or tort in a claim. When passage of time is an issue, it is likely to be a contract claim for breach of contract. When the issue is remedy, such as when an injury happens, the court will allow more expansive remedies through tort (for example, damage awards for physical injury, disability or emotional distress).

PROFESSIONAL LIABILITY

Why Should You Have Professional Liability Insurance?

Doesn't a corporate structure or placing assets in another persons name protect you and your firm? You and your firm need to have professional liability insurance for several reasons. A corporate structure or a limited liability partnership or corporation may protect your personal assets from claims in a general liability suit, but it is unlikely any professional can escape personal liability for their actions in performing their professional duty. Attempts to involve a professional in a suit even though there is a corporate structure is known as "piercing the corporate veil." Therefore, the reason for having professional liability insurance is to protect the individual practitioners in a firm and provide a means for compensating those who claim to have been injured or wronged. Notice that I didn't say to protect the individual practitioner who has committed a negligent act. The reason for this is that many times you and your insurer will elect to pay a settlement of a claim for economic reasons and not as an admission of negligence.

An Example

I'll give you an example from experience in my firm. We've been in business since 1974 and have had professional liability insurance almost since the beginning. In addition, we've had very few claims making our firm a very good investment for our insurance company. One of the first claims against us happened many years ago when a client, for whom we designed a large septic system, sued me as an individual even though we were a corporation. The first thing our attorney did was to ask for a hearing before a judge. She argued that although I may be negligent in providing faulty design of the septic system, (to be determined in court later) the claim should be shifted to our corporation since that is where the professional liability insurance was held. It didn't take much to convince the judge, since the value of our insurance was many times more than the value of my personal assets at the time. The judge ruled that, although I might be individually negligent, if the client was harmed as claimed, his best recourse was through our professional liability insurance. The case went on for several years with claims and counterclaims, hiring experts, and depositions, but it finally was settled in mediation. We made the decision that although I felt I was not negligent and the client hadn't been harmed in any way, it was best to settle the case and move on rather than spend another year preparing for a trial. Our cost to settle this case included the amount of our deductible, investments in experts on our side, and a large investment of my own time, as well as the amount the insurance company paid.

In today's world, professional liability trials almost always contain claims of negligence, breach of contract, and other claims, but usually they are really about money. Generally it is a dispute over fees or a change order for which the client doesn't think they should pay the contractor. Some professional liability insurers will pay up to the amount of your deductible if you can settle these claims without involving attorneys and the legal system. We've been successful with this method several times, and although in each case we don't believe we've done anything negligent, it resolves the claim quickly, and each party can go on rather than spend time, money, and effort preparing for a big court fight.

Another reason for having professional liability insurance is that many large clients require it. If you don't have it, you may be missing out on an important market segment. In our case, almost all of our municipal, state, and federal clients require a certificate of professional liability insurance upon signing the contract. Some even require a specific amount of coverage.

A Real Professional Liability Example of Paying Even If You Have No Fault

Finally, here's an example of how having professional liability insurance can hurt you, even if you have little to do with the case and are not found negligent. It's called the legal theory of "deep pockets."

A few years ago an automobile accident at a recently constructed intersection rendered the driver of a car paralyzed from the waste down. His medical bills and the cost of his recovery were beyond the coverage of his insurance, and he was unable to work. The suit included the contractor who constructed the new intersection and the engineering firm who designed it. To make a long story short, the contractor had gone out of business and had no assets or insurance that could contribute, so the engineering firm and their professional liability insurer were "the last men left standing," and the firm's insurer paid a claim of many millions of dollars even though there was no finding of negligence. This is an example of a case where an injury resulted in such a large lose that the court looked for a "deep pocket," in this case the engineering firm's insurance in an attempt to compensate the injured driver for his huge lose.

A Checklist of Important Items to Consider in Your Contract

There are several items to consider that may help control your professional liability exposure. It can't be eliminated but reviewing the following items can help.

1. *The agreement.* Always use an industry standard form of agreement if possible. Review client generated agreements very carefully. When in

doubt have them reviewed by your insurance provider. Be sure that your agreement contains a detailed scope of the work, a schedule for performance of the tasks described, the amount of retainer fee requested, and a clause that allows you to renegotiate if the contract is not accepted within a reasonable amount of time, say 30 days.

2. *Scope of service.* Always include a detailed scope of services, which discusses what you will and will not do. (See attachment) It should include all assumptions upon which the scope is based. These may include owner furnished items such a reasonable access to the site, a survey base map or "as-built" drawings, geotechnical investigation, and proof that the project is adequately financed. In our firm, large projects include a detailed scope, which may be several pages long. It is attached to our standard letter agreement as "Exhibit A." Small or simple projects usually have their scope directly within the standard letter agreement.
3. *Schedule.* Most clients require the agreement to include a schedule for delivery of various phases of the project or the final work product. If possible, be vague about the schedule using terms like "within 2–4 weeks or according to the client's needs," rather than be tied to a specific delivery date. As previously stated, failure to deliver according to schedule may be considered a breach of contract.
4. *Limit the length of time the proposal is valid.* Often you will submit an agreement to a client who decides to delay the work for one reason or another. A considerable amount of time may lapse before the client finally decides to begin the work and things like our availability or fee schedule may have changed. Always state in your proposal that it is only effective for a limited period like 30 days.
5. *General terms and conditions.* In addition to the contract agreement, you should provide general terms and conditions that cover items such as allocation of risk, terms of payment, access to the site, hazardous conditions, suspension of services, and dispute resolution. Our firm's general terms and conditions are copies onto the back of our agreement. This way a client cannot claim that they never saw them and, therefore, are not bound by them. Always use industry standard terms and conditions, which have been customized, with the approval of your professional liability insurer for your location and unique conditions.
6. *Allocation of risk.* This should be included in the general terms and conditions if possible. Some attorneys are of the opinion that allocation of risk is not legal or enforceable. We include a clause provided by our professional liability insurer, which they believe, is legal in our state. It is included in the terms and conditions of all of our contracts unless the client specifically asks that it be eliminated. This is the case with most

GENERAL PROVISIONS

(*Terms and Conditions*)
ACME Engineers & Surveyors, Inc. shall perform the services outlined in this agreement for the stated fee.

Access to Site
Unless otherwise stated, ACME will have access to the site for activities necessary for the performance of the services. ACME will take precautions to minimize damage due to these activities, but has not included in the fee the cost of restoration of any resulting damage.

Fee
The total fee, except when stated as a lump sum, shall be understood to be an estimate, based upon Scope of Services, and shall not be exceeded by more than 10 percent without written approval of the Client. Where the fee arrangement is to be on an hourly basis, the rates shall be in accordance with our latest fee schedule. Reimbursable expenses shall be billed to the Client at actual cost plus 15 percent.

Billings/Payments
Invoices will be submitted monthly for services and are due when rendered. Invoice shall be considered PAST DUE if not paid within 30 days after the invoice date and ACME may, without waiving any claim against the Client and without liability whatsoever to the Client, terminate the performance of the service. Retainers shall be credited on the final invoice. A monthly service charge of 1.5% of the unpaid balance (18% true annual rate) will be added to PAST DUE accounts. In the event any portion or all of an account remains unpaid 90 days after billing, the Client shall pay all costs of collection, including reasonable attorney's fees.

Indemnifications
The Client agrees, to the fullest extent permitted by law, to indemnify and hold ACME harmless from any damage, liability, or cost (including reasonable attorney's fees and costs of defense) to the extent caused by the Client's negligent acts, errors, or omissions and those of his or her contractors, subcontractors, or consultants or anyone for whom the Client is legally liable, and arising from the project that is the subject of this agreement.

Risk Allocation
In recognition of the relative risks, rewards, and benefits of the project to both the Client and ACME, the risks have been allocated so that the Client agrees that, to the fullest extent permitted by law, ACME's total liability to the Client, for any and all injuries, claims, losses, expenses, damages, or claim expenses arising out of this agreement, from any cause or causes, shall not exceed the

total amount of $50,000 or the amount of ACME's fee (whichever is greater). Such causes include, but are not limited to, ACME's negligence, errors, omissions, strict liability, and breach of contract.

Termination of Services
This agreement may be terminated by the Client or ACME should the other fail to perform his obligations hereunder. In the event of termination, the Client shall pay ACME for all services rendered to date of termination, all reimbursable expenses, and reimbursable termination expenses.

Ownership of Services
The Client acknowledges ACME's documents as instruments of professional service. Nevertheless, the plans and specifications prepared under this Agreement shall become the property of the Client upon completion of the work and payment in full of all monies due to ACME. The Client shall not reuse or make any modifications to the documents without the prior written authorization of ACME. Documents include, but are not limited to all information transferred to the Client such as CADD files, reproducible drawings, reports, etc. The Client agrees, to the fullest extent permitted by law, to indemnify and hold ACME harmless from any claim, liability or cost arising or allegedly arising out of any unauthorized reuse or modification of the documents by the Client or any person or entity that acquires or obtains the documents from or through the Client without written authorization of ACME.

Applicable Law
Unless otherwise specified, this agreement shall be governed by the laws of the State of New Hampshire.

Claims & Disputes
Claims, disputes, or other matters arising out of this agreement or the breach thereof shall be subject to and decided by Small Claims Court for amounts up to $5000 and arbitration in accordance with the Construction Industry Arbitration Rules of the American Arbitration Association for all other claims.

Pollution Exclusion
The client understands that some services requested to be performed by ACME may involve uninsurable activities relating to the presence or potential presence of hazardous substances.

Additional Services
Additional services are those services not specifically included in the scope of services stated in the agreement. ACME will notify the Client of any significant change in scope, which will be considered additional services. The Client agrees to pay ACME for any additional services on an hourly basis in accordance with our latest fee schedule.

state and federal agencies. I believe that an allocation of risk clause is very reasonable, since the amounts of our fees are very low for the potential liability we assume on most projects.

7. *Construction administration*. This is another way of minimizing or at least controlling your professional liability exposure. If your firm is retained for construction services, you are given an opportunity to impress upon your client the importance of adequate drawings and specifications and to see that the design intent is followed through. Many firms are engaged to provide permit-level drawings, and once the permits for a project are received the owner no longer calls the engineer or surveyor. These drawings end up being used for construction drawings when they are not complete nor were they intended to be used for this purpose. Beware of a client who only engages your firm to prepare permit-level drawings and does not engage your firm for construction services.
8. *Project funding*. As previously stated, be sure the project is adequately funded. If possible work with the client early on to establish the project budget. Being involved in a project, which is inadequately funded, is a recipe for disaster. Payments to you and the contractor will be delayed, and the contractor may make claims for the delays or even leave the project. Our firm has been the "no interest" banker for a project, which was in adequately funded for several years. This is not a good position to be in.
9. *The selection process*. Try to be selected on the basis of qualifications (QBS) rather than on the basis of bidding the lowest fee. A requirement to bid for work means that the low bidder also has the narrowest scope of work and must continuously ask the client for extra fee or contract amendments when any work, which is outside of the initial scope, is requested. While amendments are inevitable in all contracts, continually requesting amendments for additional fee destroys the trust relationship between the professional and the client.

I've presented a broad scope and very basic overview of the legal system and several legal items that will affect the daily practice of engineers and surveyors. This is only the proverbial "tip of the iceberg." By reading this chapter, you may have become intrigued with some of the issues and topics enough to explore them deeper. I have deliberately left out references to specific precedent setting legal cases, since they are sited in detail and delivered with much discussion in legal texts, which are readily available and referenced.

9

MARKETING PROFESSIONAL SERVICES

INTRODUCTION

At ACME Engineering & Surveying, there is more work with several residential land development projects than they can complete within the next year. The residential development market has been their "bread and butter" for the last 10 years. ACME has worked with four large developers for much of their work and has never had to compete with another firm for new work. Suddenly, the home mortgage industry has turned bad and people are not purchasing new homes. Work at ACME slows down to the point that some layoffs are necessary in order to survive. What did ACME do wrong and what is ACME to do? For starters they had no marketing plan and relied on work to come through the door the same as it had for 10 years. They assumed that the current workload trend would last well into the future. Second, they had "all of their eggs in one basket," the residential development sector and only a few clients. Third, they didn't look beyond the current backlog to estimate future work. No matter what your backlog is or how busy you currently are, you should be marketing for the future.

In our 34 years in business, we've survived several economic downturns by working in diversified market segments with a wide variety of clients. It has proven that when one market sector is down another is up, and we reorganized to take advantage of this. Let's look at how ACME can develop a marketing strategy and plan to better prepare for the next economic downturn.

Marketing professional services is a relatively new field and it's generally not done very well. A large amount of time, effort, and money can be spent on marketing efforts that either have little chance of success or cost more to market to get the project than the potential return (profit). More time and effort should be spent in market planning and developing relationships than chasing projects. If you are hearing about a project in the legal advertisements of the newspaper for the first time, it is too late. Marketing should be done at different levels by everyone in the firm, not just principals and project managers.

HOW DO SURVEYORS AND ENGINEERS GET WORK?

Prior to the Supreme Court's anticompetitive ruling most work came by word of mouth, from prior experience, or from the recommendations of others. A firm's reputation was paramount. Engineers and surveyors worked hard to gain clients' confidence and trust and then to do a good job so that they would get future projects from the client and be recommended to others by that client.

An engineer or surveyor worked very hard to earn the client's trust by providing high-quality technical designs that were cost-effective and constructible. They also provided excellent client service for a fair fee. Existing clients, if they were happy with the engineer or surveyor's work, would recommend them to other potential clients. This could work the other way too. Prior to the Supreme Court's ruling, codes of ethics forbade attempts to supplant a fellow engineer or to compete for work on the basis of fee bidding.

After the Supreme Court's anti-competitive ruling, the model for obtaining work changed dramatically. Surveyors and engineers now could bid against each other for work. Many firms were not sure how to get new clients or how to demonstrate to a potential client that their firm's work was technically very good and that they offered high quality and good client service. This was the beginning of the marketing era, not only for the engineering and surveying professions but for other professions as well.

The Supreme Court struck down the provision forbidding bidding in the NSPE Code of Ethics in a decision issued April 25, 1978. The case, 181 U.S. App. D.C. 41,47,555 F. 2d 978,984, was an appeal of an earlier injunction obtained by the Justice Department in the district court of Washington, DC. The Court determined that the "no bidding" provision of the NSPE Code of Ethics was in conflict with the Sherman Act, a widely applied federal law that was used successfully to prosecute many anticompetitive groups. NSPE's primary argument was that bidding engineering services would require engineers to commit a minimum amount of time in order to obtain a bid and, therefore, not present the owner with the most economical solution and would endanger the public safety. The Court's decision only struck the provision from the Code of Ethics, it did not *require* engineers to bid for work; in fact, the federal Brooks Law requires federal agencies to rank and select architects, engineers, and surveyors on the basis of qualifications before negotiating a fee.

Engineers and surveyors still have to work to earn a client's trust, impress them with their qualifications and perform the best client service. This is very hard to do in a price competitive environment, hence the need to market our services in order to attempt to differentiate one firm from another.

Many believe that price competition for services has reduced professional services to a commodity in the case of routine project types—that is, typical plot plan surveys, small subdivisions, small street and drainage projects, piping projects, and so forth. If the potential client asks several similarly qualified firms to submit a bid price, it is a commodity services they are seeking.

Today, firms often work to displace their competition as part of their marketing program. This is particularly true for new start-up firms. Often the approach is indirect. If a firm knows a potential client is using another firm they often approach the potential client by saying, "We understand that you are using EMCA engineering and surveying and that you have been for quite a while. We think EMCA is a fine firm, and you should be satisfied with their work, but if they are not delivering work on time and within your budget, we at ACME Engineering & Surveying would like a chance to be considered for your next project." This approach may sour some firms on their competition, but it now is completely ethical. A word of caution, however, needs to be heeded when promoting your firm. When marketing for new work in new markets, remember that most state licensure boards and codes of ethics forbid engineers from working outside of their area of expertise. If you are trying to enter a market area where you have little or no experience, team with an experienced firm, hire an experienced person for your staff, or don't try to penetrate the market.

THE SELECTION PROCESS—DIFFERENT STROKES FOR DIFFERENT FOLKS

Private Clients

Private clients can, and still do, select their engineers and surveyors on the basis of trust, experience, reputation, past performance, understanding of the project type, schedule, scope, and fee. In most cases, it is not a commodity, low-price competition. Some private clients give all of their work to an engineering and surveying firm that they've been working with for years. Other private clients deliberately split their work between two or more firms. This occurs for several reasons. First, they want to "keep them honest" in pricing the work. Second, they may have more work than one firm can handle alone, and sometimes the work is of a specialty nature, which is above the ability and experience of a general-practice firm. In most cases, private clients select their engineers and surveyors on a combination of qualifications, price, scope, schedule, and fee, using a method that works for them. The firm selected often is not the lowest-priced firm. Qualifications and experience are generally very important to private clients.

Public Clients—Federal Agencies

As previously mentioned, federal agencies are governed in procuring professional services for architectural, engineering, and surveying by the Brooks Act, Federal Law 92-582, enacted in 1972, authored by Senator Jack Brooks of Texas. It also is known as the Qualifications Based Selection or QBS law/process and has a great impact on how firms market and obtain work from federal agencies.

The law in essence states that selection of A/E/S for federal projects or those that use federal money must be made on the basis of qualifications and experience. Statements of qualifications must be requested (RFQs) so that firms can show that they are interested and experienced in the type of work for which the services are sought. The federal agency, after reviewing the qualifications, generally ranks 3–5 "most qualified firms" as the "short list" for interviews. This step is optional, and the federal agency may begin fee negotiations with the "most qualified firm" after ranking them if they desire. The "short-listed firms" may be asked to submit a detailed technical approach or proposal that specifically addresses their approach to how they'd complete the work. This may be presented at the interview or may be evaluated without an interview. Fee for services cannot be included in the technical proposal.

Once the "most qualified firm" has been selected, the federal agency begins negotiating a fee with the first choice firm. If a satisfactory fee can't be negotiated, then the firm is discharged and the federal agency begins negotiating with the "second most qualified firm." If a satisfactory fee can't be negotiated with the second choice firm, they are discharged and the procedure continues until the federal agency reaches a successful fee negotiation. Seldom does a federal agency fail in its fee negotiation with its first choice firm.

Public Clients—State Agencies

Many states have a QBS laws similar to the federal law. They often are known as a "mini–Brooks Law." Generally, it governs state agencies in the selection of A/E/S services the same as the federal law. Even where QBS laws exist, there are varying level of support at the state agency level. This can make marketing to state agencies difficult. Agencies like the DOT and environmental services, who procure or regularly oversee procurement of engineering and surveying services, strongly support the law and the process. Some agencies are unfamiliar with the law or purposely seek to subvert it by requesting price as part of the procurement process.

Many states have QBS coalitions made up of members of the architectural, engineering, and surveying professional associations. The QBS coalition's primary function is to educate state agencies and others who are not familiar with the state laws and the QBS process. In some cases, the QBS coalition may take an aggressive or legal approach with a state agency that continually ignores the law.

Public Clients—Municipalities

In states where projects utilize federal or state funds and where a state mini–Brooks Law exists, it is generally ruled that the QBS process must be followed—that is, sewer and water projects, schools, public housing, bridge projects, and federal grant projects, at the municipal level, utilizing the Transportation Enhancement (TE), Economic Development Administration (EDA), United States Rural Development Administration (USRDA), Environmental Protection Administration (EPA), Federal Highway Administration (FHWA), Federal Aviation Administration (FAA). If there are no state or federal funds included in a project and the state mini–Brooks Law doesn't control A/E/S selection down to the municipal level, the municipality can select its consultant in any manner it wants. In most cases, this results in some sort of RFP advertisement and "bid" process for engineering and surveying services. Qualifications may or may not be part of the process. I've had experience with the "two-envelope" system, where qualifications are requested and the fee is also requested but in a separate envelope. My experience has been that once the qualifications have been reviewed, it is determined that all of the submitting firms are qualified and then all the envelopes are opened. This essentially creates the illusion that a qualification-based selection has taken place when really only fee was used.

In order for the low-bid process to be successful, it *must* be based on a well-defined and often narrow scope of work. Firms that submit a bid must narrowly interpret the scope in order to have the fewest hours for any particular task and be successful in obtaining the work. This process generally sets a municipality up for many requests for contract amendments, since any work outside of the original scope needs to be paid for in order for the engineer or surveyor to make a profit. The process of asking for additional fee is common, since lay officials often write the scope for the bid before the project parameters or even the problem is well defined. This also can lead to destruction of the trust relationship between the client and the engineer or surveyor.

WHAT'S THE DIFFERENCE BETWEEN MARKETING, ADVERTISING, AND SALES?

Marketing

Marketing is considered to be the overall process of research, planning, and creating the process that will bring future work into the firm. It utilizes market analysis, research, strategic planning, advertising, and sales, each a different part of the whole marketing process. The entire process is measured and controlled by metrics that are identified by the firm. An effective marketing plan is necessary for the long-term growth and prosperity of any firm. It has been said that:

> Marketing is like a youngster in love. They are deeply in love with someone who doesn't even know they exist.

Marketing doesn't require a firm to have a marketing manager, although larger firms often do. Nor does it have to be a long drawn-out process that requires outside expertise and a lot of expense. Marketing is everyone's job, not just the principals of the firm or the marketing director. While everyone in the firm is responsible, each position has a different level of marketing responsibility. The principals provide market planning and strategic direction. They promote the firm through social contacts, providing service to existing clients, development of new clients, and overall guide the process and promote the reputation of the firm whenever possible. In addition, they oversee the proposal preparation process and usually are responsible for sales or "closing the deal." Project managers provide first class client service and maintenance marketing with existing clients. They also participate with principals in developing new clients and proposal preparation in terms of helping with scope of work and developing budgets for the work. They also should mentor younger staff on their role in marketing. Staff helps in providing first class client service and a lasting professional impression for the firm. They also help provide a high-quality work product, on time and within budget. Staff are often the first people clients come in contact with and the importance of first impression and its impact of the reputation of the firm needs to be stressed to them. In addition, some staff may be responsible for preparing project sheets for brochures, staff resumes, qualifications packages, and trade show exhibits. Years ago I even had a young staff person who constantly brought new projects into the firm both through maintenance marketing and prospecting. This person was mature beyond his years and just "got it." He now is the manager of a very large expansion of an Ivy League university.

Some firms take a dim view of marketing, since it generally involves a commitment of nonbillable time and high costs for collateral material. There is no doubt that marketing can involve a significant investment, but as you saw in the ACME example at the beginning of the chapter, it is a necessary function if the firm is going to succeed and prosper. Generally, prospecting and client development time spent establishing new relationships or nurturing existing ones is not recoverable, but time and expense spent preparing proposals can be recovered by building it into the fee for the project.

Advertising

Advertising includes all of the collateral material used to promote the firm. This generally includes the following:

- *Printed brochures.* These are three or four color glossy paper brochures that display project photos and generally present the experience and qualifications of the firm. They usually contain the firm's mission statement and "fluff" statements about client service and high-quality work product. They also may contain the resumes of principals and key staff. They are presented as a handout to prospective clients and can be quite expensive to produce. Their effectiveness or return for the cost has been questioned, since once they are reviewed by the client, they may be filed never to be seen again. They also are quickly outdated by changes in staff and recently completed projects. The general firm experience is now best presented on the firm's web site, which can be easily updated and is much less expensive. If your firm has a web site, it should be kept up to date in order to keep viewers coming back. A web site that is static is as bad as an outdated firm brochure. If you feel that your firm must have handout material or a brochure, it is best done as individual project sheets and resumes, which can be compiled and placed in a fancy folder or bound for a specific client.

- *Display advertising*. There are two types of display advertisements. Most common display advertisement is a business card type placed in newsletters, programs, and other media printed by professional and civic organizations and nonprofits. Often this type of advertising is done to support the nonprofit organization and not really for the opportunity to attract a target audience. There is little data available on the effectiveness of this type of advertising, although it does develop goodwill and enhances the reputation of the firm as a good "public citizen." The second type of display advertising is promotional adds in professional and trade magazines or other publications specifically intended to promote the firm to prospective clients. This type of advertising can be expensive, depending on the size of the advertisement, which can vary from a business card to a multi-page spread. While this type of advertising has wide circulation, it still may be difficult to measure the return in proportion to the cost. A careful consideration of advertising in publications is whether or not you are reaching an audience of perspective clients. Sometimes you may be just advertising to your competition.
- *Public TV or radio*. This is similar to display advertising placed with nonprofit organizations. Rather than being a 30-second commercial or music based sound bite, they have a very professional format announcing, "support of XYZ Public Radio is provided by ACME Surveying and Engineering, which has provided engineering and surveying to clients throughout Wonderland since 1935." Firms that financially support public radio or TV get exposure over the entire listener or viewer area, which generally is large. This type of advertising is also very expensive, but my experience is that it enhances the reputation of the firm as a good "public citizen."
- *Give-away items or freebees*. This is material is given to existing and prospective clients (and even staff) in order to help keep the firm's name foremost in their mind and that of others who see it. In the retail world, this is known as "branding." The goal is to have your firm's name become synonymous with the service the prospective client seeks. It generally consists of notepads, baseball caps, polo or tee-shirts, pens or pencils, or other such items. The challenge here is that everybody does it and to make your give-away item unique takes a lot of imagination and creativity. An architectural firm that I know does this very well. One year they gave away hand-powered generator flashlights, and another year they gave away a geometric mind teaser puzzle. Quality of the give-away item is also important, since a cheap item reflects negatively on your firm. These items also are costly.

The bottom line with all advertising is that it can be very expensive, and it can be difficult to measure the return on your firm's investment. Set a budget for advertising, and stick to it.

Sales

Sales is the process of putting all of your marketing and advertising effort together and closing the deal. It often is associated with presenting or negotiating the scope, schedule, and fee for a specific project and converting it into an actual contract agreement. Principals of firms are generally the ones to complete the sale for larger projects, although sales for smaller contracts and existing clients are often done by project managers using a standard form of agreement. Not everyone can negotiate fee and terms and conditions of a contract and close the sale. The responsibility generally falls on one or two principals and needs to be done effectively in order to ensure a continuing stream of work and profitability.

THE MARKET PLANNING PROCESS

Planning is the process of assessing where the firm is now, determining where you want it to be at some point in the future, and figuring out how to get there. Generally, market planning is done by the firm's principals and starts with research, a "brainstorming" session and an analysis of the firm's strengths, weaknesses, opportunities, and threats, more commonly known as a SWOT analysis. It can be led by one of the principals or an outside facilitator. The planning process takes time and should include input from staff and others who may not be directly involved in the planning process but whose input is valuable. Remember, if the planning process is to be successful, it needs the support of the entire staff, not just those who worked to produce the plan. Let's look at the various aspects of market planning.

Research and Analyzing Where the Firm Is Now

This includes determining what the firm does well, what the firm's areas of expertise are, who the firm's clients are, what the firm's area of geographic influence is and who the firm's competitors are. Prior to the first planning session, the participants should prepare by filling out a questionnaire, which helps them take an objective look at the firm both internally and externally. David Stone in his excellent book, entitled *Marketing in the 21st Century for Design Professionals* (ACEC, 2002) suggests such a questionnaire, and a condensed version follows.

Questionnaire to Prepare for SWOT Analysis

1. What services does our firm sell?

2. What sets us apart from our competitors?

3. Why do people pay more for our services?

4. Who are our competitors?

5. What geographic area do we serve?

6. What legacy do you want to leave at the end of your career?

7. Are you a team player?

8. What are the common values and beliefs held by most members of the firm?

9. What are some of the firm's greatest weaknesses?

10. What does the competition think of our firm?

Following initial research, the first analysis is often conducted using a SWOT (strengths, weaknesses, opportunities, and threats) analysis, which is a simple way of getting the process and creative juices going. A four-box graphic is drawn on a flipchart sheet and the facilitator fills in the boxes with input from those who participate in the planning session. A SWOT analysis session generally takes 45 minutes to an hour and a half and requires a group of 4–8 people to develop good input. A good SWOT analysis involves an honest assessment of each of the attributes of the firm by those involved and serves as the basis for the rest of the market planning. Other useful information for analyzing the firm's present situation includes historic financial data for the last 5 years, a list of new and repeat clients (including the dollar volume of their projects), a breakdown of income and profit by current market segments or areas of practice and attributes of key staff, including their areas of expertise and interest. Any information that can be determined about competitive firms also is useful, especially if they appear to be more successful in market segments where you'd like to be. Be sure to classify each attribute of the analysis as either a strength, weakness, opportunity, or threat. Some basic questions to ask when *analyzing the firm's current position* are:

- Are you highly specialized or are you a general service firm?
- Do you produce a high-quality, high-priced work product or is your firm known for low-budget work?
- Do you often compete on the basis of fee alone or do you seek to combine qualifications, high quality, and higher-than-normal fees and therefore more profits?
- Honestly, do you deliver your work product in a timely way? If not, what are the road blocks?
- Do you work best as a prime or subconsultant?
- Do you have good project management? Where can improvements be made?
- Do you have adequate staff? Are your hiring practices adequate?
- Do you have all of your "eggs in one basket" within a single market segment or with a single client?
- What service do you offer that is unique and not offered by the competition?
- Do you really know who your competition is?

The basic layout of a SWOT Analysis graphic is shown in Table 9.1.

Planning where the firm wants to be in the future is much more difficult than doing research and analyzing your strengths and weaknesses and where you are now. It requires an assessment of the future of the market for engineering and surveying services and how your firm's abilities and desires mesh with them. Data for planning the future comes from many places, including population trends, local planning records, number of building permits, trends in related industries, new or pending regulations, and overall economic trends local, national, and global. Is growth in the future for your firm? Many small firms decide that they are only capable (or desire) minimal growth. This may be based on their geographic location or the shortage of available staff. In the case of our firm it is both. We are located in a geographic location that is not growing rapidly, and since we are remote from large cities, staff availability is difficult.

Table 9.1 SWOT Analysis Graphic

STRENGTHS	OPPORTUNITIES
WEAKNESSES	THREATS

If the firm decides that it wants to grow quickly, its ability to finance growth through increasing profits or increasing debt may turn out to be an even greater limiting factor. Some additional questions to ask in the planning process when *determining growth potential* are:

- Why should we grow?
- If we are highly specialized in our technical disciplines, do we need to diversify or vice versa?
- If we are a general practice firm, are our services becoming a commodity?
- Does the quality of our work equal or surpass that of our competition? Can we use it as part of a sales approach to increase market share or are we in a market where quality doesn't matter?
- Are our fees higher than the competition with no obvious increase in client service?
- Is our overhead higher than the competition's or than government clients allow? If so, how can we trim overhead?
- Are our project management skills, and systems capable of handling more work?
- Do we have enough good project managers or can we hire them or promote them from within?

- How can we attract high-quality staff to sustain our growth? Higher salaries or more benefits? What do we have to offer that makes us different and a desirable place to work?
- Would our growth be tied to a single client or market sector and be subject to its ups and downs?
- Do we, or can we, offer unique services or expertise that will allow us to grow in a particular market sector faster than our competitors?
- Do we have the ability to "change on the fly" should our plan for growth not work out?

Answers to both the SWOT analysis and determining where you want your firm to be in the future are a good start to the market planning process, but there is more. Additional *market research* may include the following:

- Internal financial statements
- Financial statements for larger publicly traded firms
- Prospective client lists
- Number of proposals written and success rate
- Assessment of client and project type or market sector
- Analysis of clients by size of fee
- Demographic, geographic, and economic data for clients
- Assessment of marketing costs versus fee and profit
- Assessment of overhead costs
- Backlog projection
- Client interviews and surveys
- National trade publications
- Competitors web sites
- Manufacturers rep feedback
- Professional, trade, and civic organizations

The next step is developing the actual marketing plan.

Developing the Marketing Plan

In my opinion, this means "having a bias toward action." Some say that planning is a worthwhile process in itself, but nothing gets implemented unless action items are developed, someone or a group is assigned the tasks to carry out, and there are metrics for determining success or failure. In other words "avoid paralysis by analysis." David Stone's book sets an excellent outline for developing a marketing plan. The format outlined here generally follows it but includes variations that I feel make it even better. The basic elements of the marketing plan are:

1. Research & SWOT analysis
2. Positioning
3. Selling
4. Implementation

As previously mentioned, some preparation should be done by the market planning team in order for the SWOT analysis to be a productive exercise. This includes basic research into facts and figures in order to have accurate input in developing the firm's strengths, weaknesses, opportunities, and threats. Using the results of even basic research into the firm's staff, markets, clients, competition, and financial situation will allow the SWOT analysis to develop along the lines of objective input based on fact rather than subjective input based on feelings and emotion.

Positioning. David Stone, in his book, says that a firm's position in the marketplace is always changing and that the marketplace is always in a state of flux. So, why plan when your plan may be obsolete before it is complete, and besides many firms seem to be doing okay without a marketing plan? Do you know how your services stack up against the competition? How do your fees compare with others'? Are the services you offer becoming commodities? Do you have to compete based on price for most of your work? Do you know who your most profitable clients and market sectors are? Do you know which market sectors and clients look profitable for the future and which will decline? Knowing the answer to these questions determines your firm's position in the marketplace. Wouldn't you like to know the answers? Positioning is a necessary part of market planning, and effective positioning can help control how your firm heads into the future by adjusting to deal with change and each of these questions.

David Stone says there are four key questions to be answered:

1. What is it that we have to sell?
2. Who do we sell it to?
3. What sets us apart from the competition?
4. Why would anyone pay a premium for our services?

Both internal and external analysis are required here. Answers to question #1 and #2 will tend to come from internal data, while answers to question #3 and #4 will require external information. I recommend that each of these questions be answered individually (as best as possible) by members of the marketing team before they are discussed by the group. The outcome will help determine the firm's marketing mission. When developing the marketing mission, keep in mind that no firm can "be all things to all people." By the way, the marketing mission should be in agreement with the firm's mission statement.

An important part of positioning is to identify your target markets. This should be done by utilizing a decision matrix. On the left side, list the markets that your firm currently serves or would like to serve. Across the top, list the criteria that you will use to determine the requirements to enter each of the markets. Stone recommends using a score of 0–5 with 5 being the best opportunity. The results will help you better understand the markets and develop your marketing goals and action plans. David Stone's marketing matrix is reproduced here with permission (Table 9.2).

Selling the firm. Previously, I mentioned that selling is closing the deal. While in the narrowest sense this is true, selling the firm includes a lot more things, which lead to the final sale or closing the deal. Selling in the bigger picture includes tracking leads, making go/no-go decisions, writing proposals, and making presentations, whether through interviews with prospective clients or graphics presentations to accompany promotional material or trade show displays. Promoting the firm through the use of brochures and other collateral material also is part of selling.

Where do leads for new work come from? Almost anywhere is the short answer. You can get tips for new clients and new work from magazine and

Table 9.2 Stone's Market Decision Matrix

Viability Criteria / Markets	Current projects	Emerging market	Repeat clients	New clients	Firm experienced	Adequate staff	Competition	High profit potential	Risk level	Score
Boundary Surveys										
Residential Subdiv										
Topo base mapping										
Commun Towers										
Mun Sewer & Water										
Municipal Bridges										
Schools										
Recreation Trails										
Comm Septic System										

Score each market by desirability of market
0 = undesirable
5 = very desirable

newspaper articles, legal advertisements, friends and relatives, suppliers, existing clients, research data on clients needs, and unsolicited inquires. Once you've received a lead for a new project, you need to make a go/no-go decision. Since often there are more opportunities than a firm can handle, the go/no-go decision is key to narrowing down the opportunities to those that have the greatest chance of success. A sample go/no-go decision form follows.

ACME Engineers & Surveyors, Inc.
GO NO/GO PROBABILITY FACTOR CHECKLIST

Date:________ **Dept:**______ **Preparer Initials: :**_______ **Proposal Due Date:** ________

Prospective Project Name:____________________ **Prospective Client:**__________

Project Location:____________ **Source of Lead**____________**(Attach RFP, RFQ or Ad)**

1. Relationship:	**Rank: (circle choice)**
Existing favorable client	100%
Past client w/good reputation	75%
Team member relationship	50%
Potential client w/prior contact	40%
Potential for repeat work	35%
Referral	30%
Cold call	10%
Past client – poor relationship	5%
2. Similar Experience:	
5+ Recent projects	100%
4+ Recent projects	75%
3+ Recent projects	50%
2+ Recent or old projects	20%
1 Recent project	10%
0 Recent projects	0%
3. Competition:	
0 Firms competing	110%
2 Firms competing	100%
3 Firms competing	33%
4 Firms competing	25%
5 Firms competing	20%
More than 5 firms	10%
4. Geography:	
Within 75 miles of office	100%
Within 100 miles of office	75%
Within State (NH & ME)	50%
Outside of State	25%
5. Marketing Effort:	
0-1% of possible fee	100%
2-3% of possible fee	60%
4-5% of possible fee	35%
6+% of possible fee	0%

6. Price Importance:	**Rank:**
QBS process (negotiation)	100%
Separate cost proposal (two envelope)	80%
Moderate importance	60%
Important	40%
Primary consideration	10%
7. Access to Decision Makers:	
Meet or know all decision makers	100%
Meet 1 or know a decision maker	80%
Meet a decision influencer	60%
Meet w/procurement officer	40%
No contact/formal RFP or Response to newspaper ad	10%
8. Ability to Perform:	
Resources available immediately	100%
Resources will be available shortly	80%
Resources won't be available soon	50%
Resources won't be available	0%

COMPUTATIONS:

1. Relationship	_____
2. Experience	_____
3. Competition	_____
4. Geography	_____
5. Marketing Effort	_____
6. Price Importance	_____
7. Access to Decision Makers	_____
8. Ability to Perform	_____
Total:	_____
Total divided by 8	_____%

__ **Proposal Cost Estimate:**

1. Estimated project fee:	$_______
2. Estimated proposal preparation cost:	$_______
3. Reasonable cost vs. expected profit:	$_______

GO____ **NO GO**______ **BY**______

Once you've made the decision to pursue a project, tracking the lead begins. The next step is doing more homework in order to better understand the client's problem so that a preliminary scope can be developed. This may include face to face meetings with the prospective client (be sure that you know who the decision makers are), site visits, and review of background material. Remember, long-term relationships are what you should strive for not a single project. The details of proposal writing and making presentations will be covered later.

Advertising material includes brochures, display advertising, give-aways, and public radio and TV sponsorships previously mentioned. Selling the firm includes almost anything used to promote name recognition and branding. Beyond advertising, promoting the firm may include press releases, newsletters, web sites, awards, and trade show exhibits.

Press releases should be written for any major newsworthy event involving the firm. A good example is the firm winning an award or the addition of new partners or ownership changes. An important thing to remember about press releases is that they may or may not be printed; it's the editor's choice. Get to know the editor of the publications where you'd like press releases to be published and, more importantly, prepare press releases according the publication's style guidelines.

Newsletters take a lot of time and staff commitment. The common scenario is that a firm gets enthused about printing a newsletter, key staff participate in writing the stories, and after a few issues it dies a slow death because of lack of time and waning enthusiasm. The most successful newsletters not only "blow the horn" of the firm with articles about new clients, contracts, and awards but also contain humor and helpful tips for readers. It must contain something to make the reader look forward to the next edition.

Web sites are an inexpensive method for distributing information and promoting the firm to the world. A web site often contains several sections, which can be reached by clicking icons or topic headings on the "home" page. The home page itself should be clean and simple and entice the visitor to explore further. Wed sites can contain job advertisements as well as project experience, resumes, and contact information. If you have a web site, it must be at least as good as your competition. It is better not to have a web site than to have a poor one or one that reads "under development" when you go to the site.

Participating in trade shows is an opportunity for your firm to promote its qualifications to a target audience who purchases engineering and surveying services. Display kiosks for trade shows are very expensive and need constant updating, since last year's material is outdated and may project a negative image if it is viewed by the same prospects who viewed it a year ago. Most of your competitors will be displaying at the same trade show, so direct comparison of a firm's collateral material and give-aways is easy. The return on your investment from trade shows is also difficult to measure. Often attendees at trade shows represent the prospective clients you seek, but those attending are not the organization's decision makers.

Implementation

Now that most of the work has been done the only thing left is summarizing all of the work into the plan document and implementing it. Setting goals and objectives should be done by the firm's principals in conjunction with the marketing team. It should include input from senior staff. The plan must be put forward by the firm's principals; otherwise, there isn't demonstrated leadership to implement the plan. If input isn't sought from others, they will insist that it is management's plan and will not share in ownership of the plan, making implementation difficult.

Elements of the marketing plan. Most marketing plans start by repeating the firm's mission statement from the company's strategic plan. The overall objectives of where the firm would like to be in 1, 3, and 5 years are usually strategic and those related to marketing should be stated. The strategic objective may be something like "our firm plans to be the dominant firm furnishing general engineering and surveying services to private, municipal, and government clients in our geographic area." Marketing goals that support the strategies should be next. For example, "we must double our client base, staff, and dollar volume within 10 years in order to be (or remain) dominant in our area." The marketing plan should evolve into a document that is readily usable by all of the firm's management. It should contain a table of contents; an executive summary, including the mission statement; and a brief summary of the action items. The actions items should include details of each marketing goal and the actions necessary to accomplish the goal, including who is responsible, how much effort and money is to be spent, and a schedule for completion of the item. Backup material that was used to develop the plan should be included in the appendix.

The marketing plan also should contain:

1. Marketing budget, including staff time commitment and marketing material
2. Fee income & profit goals broken down by client type or market segment
3. Other growth goals, including new market segments, staff, facilities, and branch offices (the last two may be included in the strategic plan too)
4. The firm organization chart showing marketing responsibility by job or position

5. Action plans; each action item should be an action that supports a marketing goal. These may include:
   a. Establishing a schedule and budget for the action item
   b. Determining who is the lead person responsible for the item
   c. Establishing the metric for measurement of success
   d. Establishing a feedback mechanism for future action

PROSPECTING

Prospecting for new clients includes a variety of methods such as using prospective client lists, making cold calls, running seminars and doing public speaking, sending direct mail, maintaining trade association membership, and networking. Ford Harding, in his book *Creating Rainmakers* (Wiley, 2004), says that prospecting is somewhere between positioning the firm and selling. It is the process of turning prospects or leads into real clients. Harding goes on to say that a firm needs a single process that serves as the engine in a complete system to keep work coming into the firm. One person in the firm needs to be in charge of the system and delegate the work to others. A sample prospect list is shown in Table 9.3.

Prospecting processes, to develop new leads, may include:

- Article clipping from newspapers and magazines
- Cold calling
- Presenting seminars or public speaking
- Networking through relationships
- Trade association membership
- Following up on referrals

The most important thing about prospecting is taking the time to do it. Most engineers and surveyors are more comfortable with technical work than with making calls or developing relationships to bring in new work. Some even feel that its is beneath them or unprofessional to be asked to do prospecting. In short, they are not rainmakers. Most engineering and surveying firms need prospecting processes and a system that includes someone on staff, possibly a marketing assistant, to do basic research and make introductory calls to set up appointments for principals, who will attempt to establish a relationship and sell the firm's services. Once the system is working and data is available, the principals need to set aside a certain amount of time each day or each week to make calls or follow up on prospect leads. Some contacts may be cold calls and some may be to reestablish old relationships.

Cold calling itself is a process. In order to make it productive, some homework needs to be done first. The goal is to begin to develop a relationship,

Table 9.3 Prospect List

ACME ENGINEERS & SURVEYORS, INC.
PROSPECT LIST
PREPARED BY: AXLE ROD
DATE: JANUARY 1, 200X

COMPANY	CONTACT PERSON	PHONE	FAX	EMAIL	CURRENT CLIENT YES/NO	PROJECT NAME	MARKET CATEGORY	CONTACT STATUS

which will lead to you having a better understanding of the client's concerns and needs and discover how you can work together to address them. Exploring the barriers to your firm being hired is also one of the goals for a cold call. They may be that the prospective client already works with another engineering and surveying firm or that the current budget doesn't allow their projects to move ahead this year. If a cold call can result in a face-to-face meeting, you have taken the next step toward developing a long-term relationship. Cold callers must be persistent and not take rejection personally. Remember, selling is a numbers game, so the more calls made, the more chances for obtaining a new client and project.

Presenting seminars or public speaking are prospecting methods that give the person making the presentation instant credibility as an expert, which by association enhances the firm's reputation. Not all principals possess sufficient expertise or are comfortable presenting seminars or making speeches, but at least one principal in the firm should be assigned the task.

Trade association membership is also an important method of prospecting. In this case, you are speaking directly to prospective clients or those who can give you leads. In our firm, we meet prospective clients as well as existing ones each year at a municipal trade show where we have a booth, which displays examples of our engineering and surveying projects and services. Membership on any board or commission that exposes you to prospective clients is an important part of the prospecting process (see Table 9.4).

Table 9.4 Relationship Building Matrix

Relationship Technique	Flexibility	Strength	Weakness	Works Best With
Cold calling	Low	Face-to-face Contact	High Cost	Needed Service
Seminars	Moderate	Prequalifies prospects	Requires long lists	Frequent needs of firms
Publicity	High	Reaches broad audience	Lessens your control	Services with high news value
Relationship Marketing	High	Low cost	Doesn't broaden base	Frequently needed services
Direct Mail	High	Low cost	Low control	In combination with other approaches
Public Speaking	Moderate	Prequalifies prospects	Requires strong speaking ability	When many potential audiences
Trade associations	Moderate	Face-to-face meetings	Competitor will be there, too	Highly active markets

From Chapter 8 Exhibit 8.2 *Creating Rainmakers* by Ford Harding, Adams Media Corporation, 1998.

I mentioned earlier that in most cases engineering and surveying firms don't have a rainmaker as such; some further definition is needed here so that you can recognize whether or not you do have one and make the best use of such a person. Generally, a rainmaker is a principal of the firm who has a natural inclination and ability to enhance the firm's reputation and sell the firm's services. Rainmakers are never negative, even after rejection. They consistently devote a significant amount of their time to networking and developing leads while also maintaining a high level of billable time. They generally are known by the firm's clients as the "go to" person or the "fixer." If your firm has such a person, consider the firm blessed.

MAINTENANCE MARKETING & CROSS-SELLING

Maintenance marketing and cross-selling are terms not often used in engineering and surveying firms. They have to do with keeping existing clients happy by serving them well and establishing a relationship that guarantees that the firm will get the next project the client has to offer. Who's responsible for maintenance marketing and cross-selling? Everyone in the firm, but particularly project managers and project engineers. That doesn't mean that the principals or administrative assistants shouldn't constantly be aware of the client's next project or expanding a client's understanding of the full line of services that the firm has to offer. Maintenance marketing includes cross-selling, but they are different. What do I mean by this? A project manager and his team regularly provide structural design services to an architect to design the framing system for buildings. He is always aware of the next structural design project with this client, so he has an opportunity to write proposals for every structural project that the architect has coming, but he also should be educating the architect that his firm can furnish other services, including surveying, civil site design, and permitting. If he's successful in cross-selling additional services, the fee to the firm can be increased considerably. Every project manager and project engineer should be aware of maintenance marketing and cross-selling opportunities with every client.

INTERVIEWS & PRESENTATIONS

Once the firm has been short listed, the next step in the selection process is generally an interview with the client's selection committee. Sometimes, I think interviews are a "crap shoot." I'll come out of an interview thinking: we had good eye contact, the body language seemed to work, the potential client asked good sincere questions, and they kept us beyond our allocated time. I'd say to the other members of the presentation team "I think we won," only to hear later that the project was awarded to another firm. A "crap shoot!" Our firm wins about 50 percent of the projects that we interview for, so I'm no expert, but here are some of the ways we prepare for an interview presentation.

First, when we are notified that we've been selected for an interview I like to be the last firm interview if given the choice. Why? There are several reasons. If we arrive on time or a little early for the interview we may get to see who our competition interview team is and some of the material they used in their presentation and then make last minute adjustments if needed. I also like to be last because leaving the last impression with an interview committee has the opportunity to have the greatest impact.

Second, we really try to do our homework in preparation for the interview. We try to determine who our competition is. How have we faired against them in the past? What type of presentation do they generally put on, simple and straight forward or glitzy and corporate? I also like to get to know the interview committee as best as I can. Have we worked with this client in the past? Are they going to be using a score sheet to rank the firms and what is the criteria? Who is the member most likely to influence the decision? Do we have a relationship with him or her? What is the true nature of the problem and is the project straightforward or is it complex and not very well defined? Is fee going to be part of the discussion or negotiated later? What is the budget for design or is there a total project budget? What is the profit potential or the potential for repeat work? I feel that asking a lot of questions while preparing for the presentation shows the potential client that you are interested in their project and discovering their true nature of their needs.

Third, we decide which graphics are needed for the presentation. Should we repeat graphics previously presented in the qualification statement or technical proposal? Are project photos necessary? How are the graphics going to be presented? Will they be mounted on foam-core board and presented on easels or will a PowerPoint presentation make a greater impact? I generally use the following graphics:

- Team organization chart, including the client representative and consultants
- A task and responsibility matrix that details the phases and tasks as we see them and who will be responsible for which task
- A project approach flow chart that shows the steps in the process and looped "what if" decisions if several alternates are studied
- A preliminary schedule
- A preliminary budget if it can be defined or how we see the client's budget being used, if the budget has already been determined
- Miscellaneous photos of similar projects if necessary

Typical figures used by our firm for client interviews and presentations are shown in Figures 9.1 to 9.4.

Once the presentation is complete, it should be rehearsed in your office. Pick several people to be the mock client team, who will ask questions similar to what you expect the client's team to ask. Rehearse the presentation to be sure that you can do it within the allocated time and decide whether you would like to be interrupted for questions as the presentation goes along or ask that they be held until the end.

Finally, the time for the presentation has come. Be sure that the team is dressed appropriately. Since our firm is located in a rural community, overdressing for a local interview places us in the same category as the "experts from out of town," and we are conscientious of that. Generally, a sport shirt and kakis work well locally. I have a saying in the office that goes like this, "If we are going south of town (to a more urban community) we wear a tie to the presentation (since the interview committee is likely to be dressed this way too), and if we are going north of the mountains, we wear red and black checkered wool." My point being that if we are to make a good first

Town of Newville
Fairview Road Bridge over
Happy Brook

Responsibility / Task	ACME				CONSULTANTS			
	ABC	*DEF*	GHI	JKL	Wet Sc	XY Borings	Town	ABDOT
1. Project Management & Client Liaison	●	○					○	
2. Survey & Base Map	○	○	●					
3. Wetland Delineation		○			●			
4. Soil Borings		○		○		●		
5. Geotechnical Evaluation	○	●		○				
6. Hydrology/Hydraulics		○		●				
7. Eng. Study (Alternates) & Cost Estimate	○	●		○				○
8. Select Preferred Alternate	○	○					●	○
9. Preliminary Design	○	●		○			○	○
10. Final Design (PS&E)	○	●		○			○	○
11. Advertisement for Bids		●		○			○	
12. Contract Award	○	○					●	○
13. Construction Observation	○	●		○			○	○

● Primary Responsibility ○ Secondary Responsibility

ACME — ACME Engineers & Surveyors, Inc., Funtown, NH

Figure 9.1 Task & Responsibility matrix.

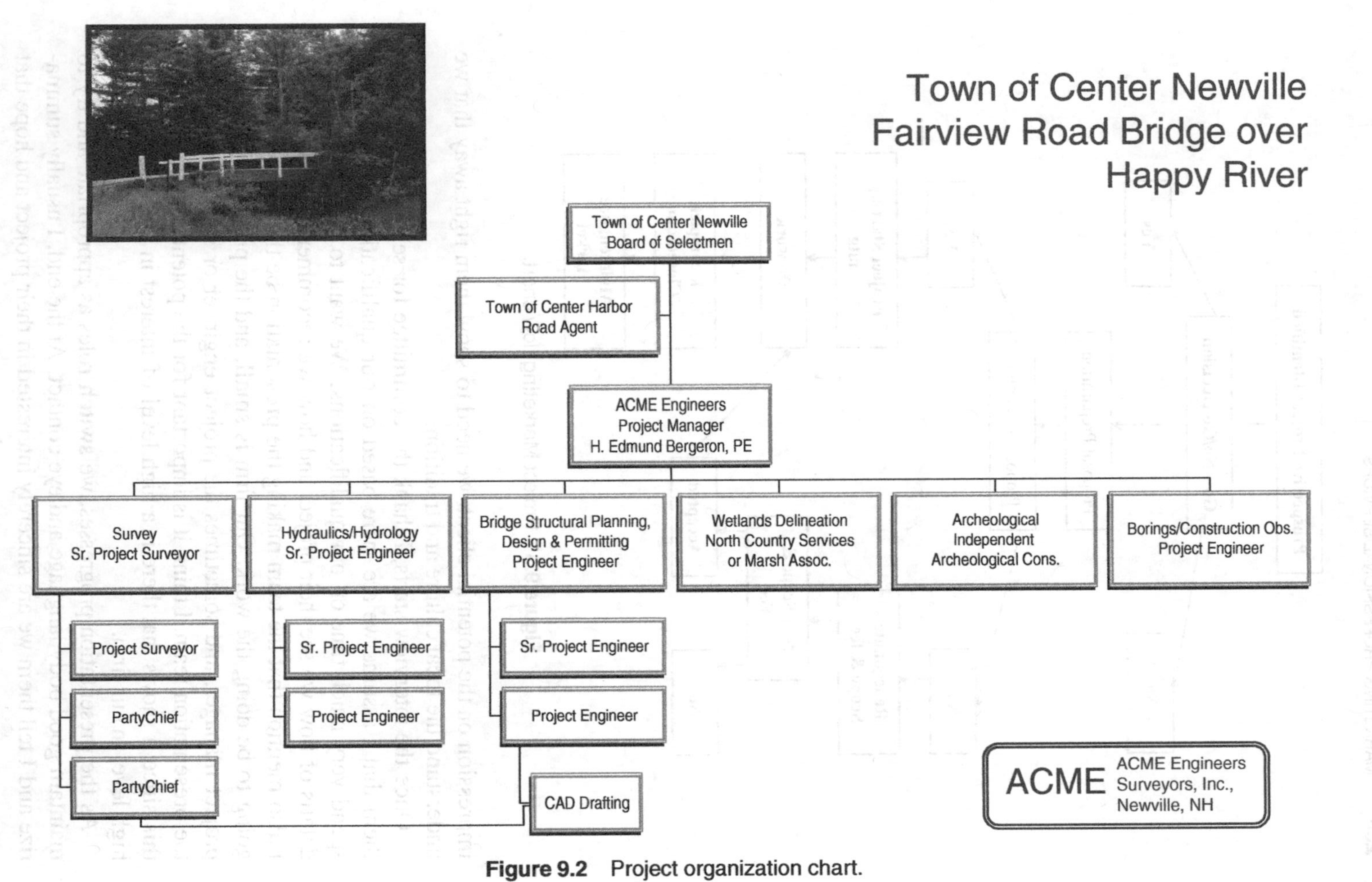

Figure 9.2 Project organization chart.

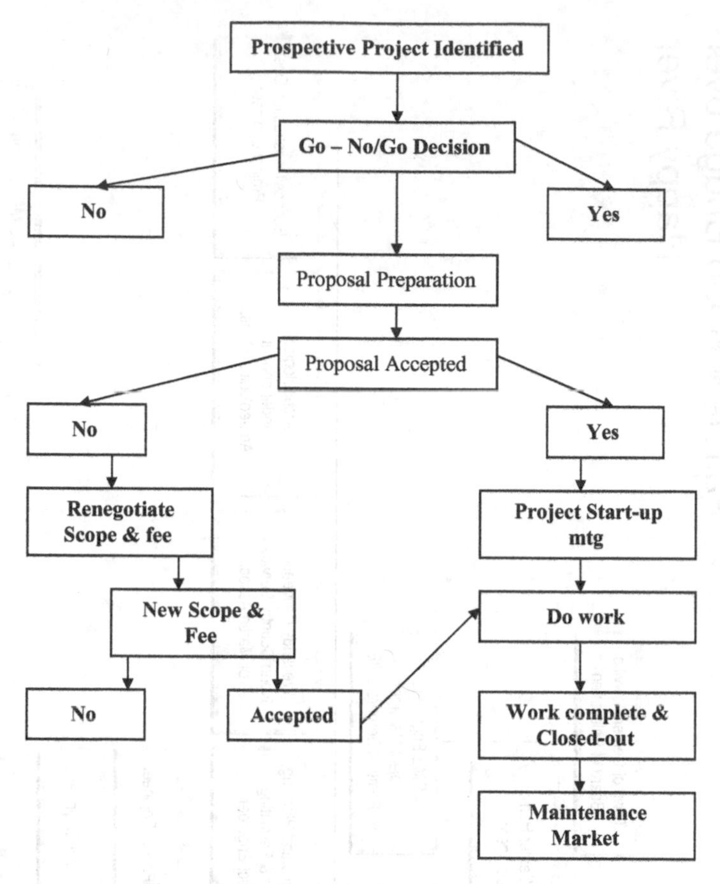

Figure 9.3 Project Marketing flow chart.

impression on the potential client, we need to show them right away that we understand the local culture and tradition.

Once the interview starts, I thank the committee for selecting us and tell them that I assume we are there based on our qualifications, so we won't spend very much time on our qualifications. We want to drill down on the details of how we see their project and how we recommend approaching it. I also mention that the team making the presentation be the people who are going to be doing the work. Our firm is small, and the principal-in-charge, project manager, and sometimes the project engineer or surveyor make up the presentation team. I think it is important for the potential client to know this since it shows that there is a high level of interest in their work from a high level in our firm.

As the presentation progresses, we switch roles as appropriate and try to maintain good body language and eye contact. At the end, I usually summarize and I tell them we are sincerely interested in their project and hope that

Town of Bethel, Maine
Sewer Storm Drain Proposal

PROJECT SCHEDULE

Month	March				April					May				June				July				
Week Starting Date	5	12	19	26	2	9	16	23	30	7	14	21	28	4	10	17	24	2	9	16	23	30
Phase & Task Description																						
Contract Authorization																						
Existing-Features Survey																						
Item 1: Design of Reconstructed Sewer																						
Data Collection																						
Preliminary Design																						
Final Design																						
Item 2: Design of New Sewer																						
Data Collection																						
Preliminary Design																						
Final Design																						
Item 3: Design of Storm Drain Improvements																						
Data Collection																						
Preliminary Design																						
Final Design																						

Figure 9.4 Project schedule.

they select our firm. At this point, I feel we've done the best we can and all there is left to do is wait. I don't predict the outcome any more; it's just a "crap shoot."

The key is doing your homework, preparation, and practice.

SCOPE OF WORK

Writing an effective scope of work is the end result of marketing. All that is left is to sign the deal. A scope for a small project is often done by a principal or project manager, while a scope for a large project that involves several disciplines and consultants may require input from several people. Writing a scope of work can be very time-consuming so it only makes sense to standardize them when possible. In our firm, we have several standard scopes for various types of civil and structural projects that we do over and over again. When this type of scope is applicable, it becomes a cut and paste or filling the blank task, which can be done rather quickly. Scopes are always attached to our standard letter agreement as "Exhibit A." The overall format for a scope of work in our firm is the same whether it is a large or small project, only the size and effort change. The format is as follows:

- *Introduction*. This is the opening that references the parties to the agreement, the date, and a statement that this scope is an exhibit and becomes part of the agreement.
- *Project Approach/Understanding*. This section may be several paragraphs long. It often repeats some or all of the client's RFQ/RFP description of the work, but it also describes the approach to how we will do the work in some detail.
- *Assumptions*. This is a statement of our understanding of items that we assume are necessary for proceeding and the completion of the work. It may include statements like, "the owner is to provide survey and geotechnical investigation" and "the owner has secured adequate financing for the work," or the owner is to provide access to the site." These generally are not deal breakers but are intended to make it clear that the owner has some responsibilities to fulfill as part of the work.
- *Scope of services*. Here is where we get into the detail of how, who, and how much it will cost to do the work. Many of our projects are multiphase, so the scope will be subdivided as follows:
  - *Pre-design phase*. This includes all work to be done by our firm before the actual design can begin. It includes research of information

regarding the project and any previous work done at the site. It describes survey base mapping, if we are providing it, as well as geotechnical investigation. Other pre-design work that must be done may include archeological and historic research and field investigations, wetland delineation, and traffic counts.

- *Study phase (schematic design).* This phase includes all of the work necessary to determine the basic and controlling design parameters as well as investigate several alternatives. This phase generally includes development of schematic plans and sketches for each alternate as well as engineer's opinion of probable project cost. A word of caution here. Be sure to include an adequate contingency in the opinions of cost, since the plans have only been developed to the schematic level, and there are still plenty of unknowns. In our firm we always use a 20 percent contingency for opinions of cost during this phase. Presentation of the study and alternates usually includes at least one meeting with the decision makers and often a public presentation if the project is sensitive or has a high level of community interest. From the study phase, a preferred alternate is either suggested by our firm or voted by the client. It is necessary in order to proceed to preliminary design.
- *Preliminary design phase.* This phase describes development of the preferred alternate to the level of 70–90 percent complete drawings and opinions of cost. Since many of the unknowns have become apparent the contingency is reduced to 15 percent during this phase. Usually, drawings from this phase contain enough detail to apply for local and state permits. This phase usually includes several meetings to obtain concurrence with development of the design.
- *Permitting phase.* This includes work related to applying for local, state, and federal permits. It may include meetings with the permit agencies and usually includes furnishing additional information to them as requested.
- *Final design phase.* This phase includes finishing the drawings to 100 percent so that they are suitable for bidding and construction. It also includes creating the project manual, which includes all of the technical specifications as well as all of the legal documents, including the invitation and instruction to bidders, general terms and conditions, supplemental conditions, and the agreement. A final engineer's opinion of cost is also prepared with the contingency reduced to 10 percent. A final design review meeting usually is held with the client.

- *Bidding phase*. During this phase the project is advertised for bids and distribute drawing and specification packages. We usually conduct a pre-bid meeting to familiarize bidders with the site and take questions regarding the drawings and specifications. These are clarified by issuing an addendum to all holders of plans and specs. We also conduct the bid opening, check that the bids meet all of the requirements such as bonds, certificates of insurance, and acknowledgment of addenda. Once this is complete, we make a recommendation to the owner for award of the contract.
- *Construction administration phase*. First, we try to be sure all of our agreements include construction administration services. This is the best way to help ensure that the design intent is followed and limit the possibility of legal issues related to the design. Construction services agreements are always included in our public works contracts but some private clients eliminate it as a cost-saving measure. I think this is "penny wise and pound foolish."

Just a word about deliverables and meetings. Our scopes of work always include what the deliverables are as well as the number of meetings that we expect to attend in the description of each phase. That way the client clearly understands what they'll be getting as the result of each phase. The number of meetings needs to be specified so that the client understands that you can't go to an infinite number of meetings without asking for additional compensation. Meetings can have a serious impact on the budget.

- *Fee statement*. Following the detailed scope of services is the fee. It is broken down into the fee for each phase of the work. The fee can be lump sum or an hourly estimate. In our firm, we always use a lump sum fee when we are familiar with the type of project and scope. It gives us a goal, and if we beat the number of hours estimated we make a bigger profit. We use hourly fee estimates for the phases or tasks that are difficult to define the level of effort or when the estimated hours to perform a task are more of a "guesstimate" than a solid estimate. Hourly fees are often used for permitting and construction administration phases of a project. A sample project fee budget is shown in Table 9.5.
- *Schedule*. This section of the scope details our schedule for doing the work. It includes specifying the expected completion date for each phase or other milestone.
- *Project team*. This is where we specific who the project team is. It should be the same project manager and project engineer who were present at the interview. Other members of the team including the project surveyor and key consultants also are specified here.

Table 9.5 Sample Project Fee Budget

Newville Fire Department No. 1 Town of Center Newville, Maine	Amount
Data Collection	$ 2,153
Preliminary Design	$ 7,386
Permits	$ 6,237
Final Plans	$ 6,046
Preliminary Structural Design	$ 10,773
Final Structural Plans	$ 7,968
Grand Total	$ 40,565

NEGOTIATING THE AGREEMENT

I am convinced that many engineering and surveying firm principals do not know how to negotiate a scope and fee that will provide for achieving the client's goals, provide high-quality technical solutions, and allow them to make a fair profit on the work. Often, when presented with a scope and fee proposal, the client's first response is, "the fee is too high" or "the fee is over our budget." The consultants response is often, "I'll review the hours and get back to you." He has just been set up by the client to provide the same level of work for less money. In essence, he has just agreed to do the work for less profit. This is not a successful negotiation, but it happens all to often in small firms.

Keep in mind that engineering and surveying firms sell man-hours. If, in preparing your initial fee proposal, you felt that it would take 100 man-hours to complete a survey, based on many years experience with similar projects, why would you think you could complete the work in fewer hours when your client responded that your fee was too high and your response was that you'd review the man-hours? Your initial estimate was probably right, especially if you solicited input or verification of the estimated man-hours from those who will actually be doing the work. So what are you to do? The first thing is to take the time to review the scope of work in detail with your client. Show him your estimated man-hours for each phase and task. Discuss what reimbursable items you have included such as mileage, consultant fees, meeting attendance, or permit application fees. Discuss the method of proposing your fee, is it lump sum or hourly? Once you've done this most clients will understand the value they are receiving for their investment in your firm's services and agree that the initial proposed fee is reasonable and fair. If they don't then the following approach is what we do in our firm.

If after a detailed explanation, the client is still insistent on reducing your fee, you must discuss what services the client can provide himself or how you can reduce the scope without reducing the quality of your services. Some firms suggest that the client engage and pay the consultants themselves. This will save them the cost of your mark-up, usually 10–15 percent of the consultant's fee. The client also can pay their own application fees, which often are substantial, but this really only shifts the cost from you to them unless your firm marks up application fee, too. Meetings are an area of potential great savings. Often a client insists on being very involved in the progress of their project, which results in a large number of meetings. Substantial savings may be possible by reducing the number of meetings in your original scope and providing a per meeting fee for additional meetings if necessary. If these items do not provide the client with a satisfactory fee, then it is likely that a reduction in scope is the only other option. Reducing the scope of work is a delicate balance between meeting the client's needs and continuing to provide high-quality work. Where do you cut and still maintain quality? For a small survey project there may not be much room. Maybe setting property corners is left out until a later date? Maybe field work under winter conditions doesn't justify the expense? Maybe the client provides some of the deed research or possibly clears some of the lines so that field work will progress more quickly? Is it possible to do the work and not provide a finished, recordable map? These are all items that may be deleted from your original scope of work in order to reach a lower fee.

In the case of a larger project where your firm has been selected on the QBS basis and now it is time to negotiate scope and fee, there is a greater opportunity for fee savings. Some of the work, such as survey base mapping and geotechnical investigation can be provided by the owner. Is it possible to limit the number of alternatives studied in the early phases of the project? Can permits be applied for with preliminary rather than final drawings? Can the amount of travel to distant meetings be reduced? Can construction phase services be eliminated, which I do not recommend. It has been my experience that most clients who select their consultants on the basis of qualifications have a pretty realistic range of what the scope and fee for a project should be when the negotiations start and quickly reach agreement with the firm with only slight adjustments. In fact, I asked a DOT chief bridge engineer if he'd ever discharged his first choice consultant and went to the second-ranked firm because a successful fee couldn't be negotiated, and he said that in his experience it had never happened. It appears as though the system works.

An important item to remember when negotiating scope and fee, especially if a significant fee reduction is the final result, is to remind your client that any requested changes in scope will require a contract amendment or additional work authorization. Many clients are skilled at negotiating a lower fee based on a reduction in scope and then asking the firm to do additional work (outside of the scope and possibly items that were negotiated out) without a corresponding fee increase.

Finally, an important part of negotiating is the billing method. Are you charging the client a lump sum fee or will he be charged hourly for your services? Often, when a scope is well defined and the firm is familiar with the type of project a lump sum fee is charged for some or all phases of the work. This gives the client a complete understanding of what his cost will be and allows the firm to improve its profit if the work is completed ahead of schedule for fewer man-hours than originally anticipated. The firm also assumes the risk of being over the budgeted man-hours and not being paid for them unless there is a significant change in scope agreed to by the client. When the scope is not well defined, firms often provide an estimated fee and then invoice for the actual hours that the work requires. Here the risk is taken by the client, and if the number of man-hours is beyond the estimate, he should pay for the work. Beware of clients who negotiate an hourly not to exceed fee, since this shifts all of the risk to the firm.

Remember the number one rule of negotiating. Determine the point at which you need to walk away, since it is no longer a fair deal for you and your firm; when you propose to do this, the client likely won't let you walk away.

Now you have it. Marketing is everything from research into who your clients are to determining your firm's strengths and weaknesses. This should result in a marketing plan that guides the firm toward achieving its goals. An important result of marketing is a good scope with a reasonable fee and then closing the deal. Finally, remember that most of our work, 70–90 percent comes from repeat clients, so the most important part of marketing is to provide first class client service to existing clients.

10

OWNERSHIP TRANSITION

INTRODUCTION

The sale of an engineering or surveying firm is a special transaction, and just as no two firms are exactly the same, no two sales are exactly the same. The sale of a firm by the founders to the next generation may be difficult, since they likely have identified their life with the firm for a long period of time. A founder's emotional attachment often has a lot to do with "the plan" for the next steps in their life. Hopefully, the new owners will find a role in the firm for the founders that will allow them to stay involved and contribute in a meaningful way.

Most firms need the help of a professional who is experienced in engineering and surveying firm transactions to be successful. We determined many years ago that our firm was different, and based on this, internal ownership transition would be the best thing for our clients and staff. About 10 years ago I began to contemplate the sale of our firm. Since it was at least 10 years until my planned retirement, I thought getting started now would allow plenty of time to identify the new owners, get them trained in leadership and management, and finance the transaction so that it would be affordable to the young staff who had little in terms of personal assets. I started with no outside help, other than an agreement drafted by our attorney, and offered to gift stock to three leaders who had been with the firm a number of years. This transaction failed for a variety of reasons but probably the single most significant one was that the three individuals just weren't ready. They were making good salaries and couldn't see what was in it for them other than increased responsibility and more work.

For our second attempt we hired a consultant who was experienced with transitions of architect and engineering firms of all sizes. My partner and I had attended several seminars and had read all of the books on ownership transition, so we thought we were ready. This time we identified a group of six people who were leaders in the firm. Several had been part of the previous attempt, so we felt that they had matured and would lead the others through the process. Our consultant interviewed each individual, determined the value of the firm and made several presentations over a period of a few months. This attempt also failed. This time it was for a different set of reasons. Several members of the group did not trust each other, and some felt the consultant's valuation of the firm was too high and the projections for future growth (based on past performance) were unrealistic. My partner and I were disappointed and felt that we had no choice but to seek other options. Our time was running out.

In the interim, we talked with a much larger firm who was interested in acquiring our firm, but the deal didn't sound very enticing. They proposed to put very little cash down and essentially buy us, over a period of years, with profits that we generated ourselves. It sounded like a good deal for them but not for us. We also came close to merging with a firm similar in size and philosophy to ours, but for a variety of reasons that one didn't work either. In each of these cases, our staff would have suffered various consequences, so we returned to the idea that internal transition was the best way to go.

Three years ago we decided that an employee stock ownership program (ESOP) would be the best way to go, so we got started. An ESOP is a trust that holds stock of the firm in each participating employee's name in proportion to an IRS formula for qualified benefit plans similar to a 401(k) plan. We continued to work with the same consultant, since the firm had to be valued and the ESOP established in accordance with IRS regulations in order to be a qualified plan. We started with me selling 30 percent of my stock to the ESOP. Each year everyone in the firm accrues stock in the ESOP, which is purchased with profits. In addition to our 401(k), this truly was another benefit for our staff. More details on ESOPs later.

THE LEADERSHIP DEVELOPMENT PLAN

I felt that finally we were headed in a direction, which could ensure the future of the firm for our staff and clients and allow my partner and me to begin planning for our retirement. I also felt that, if we were really going to be successful, we needed a solid leadership group (to replace us) who where technically superior, could provide excellent client service, could effectively manage projects, could work together as a team, was enthusiastic about the future, and possessed the leadership and management skills to follow through and be successful in our absence. Although we had a few candidates, there were no obvious standouts, so we began a leadership development program.

I felt that if the firm was going to be successful over the long term, stock had to be distributed to a leadership group outside of the ESOP. Our consultant called this "skin in the game." Over a period of 12–14 months we held leadership development seminars, after work, covering all sorts of management and leadership topics, including the basics of financial statements, project management, communications, management of self, motivation, and mentoring. The seminars were open to everyone in the firm, and participation by all was encouraged. We specifically held the seminars after hours in order to see who was motivated, committed, and interested enough to make the needed arrangements and to see who would show up consistently. I remembered a quote of Richard Weingardt's in his book, Forks in the Road (Palamar Publishing, Denver, 1998) "*The world is run by those who show up.*" After a year, five people were consistently showing up and participating at a level that demonstrated they were interested and had the ability and desire to take on more responsibility and learn more about the firm's operations. We asked each of these people if they would be interested in being stock owners, and although they weren't sure how they would pay for their stock, each indicated that he/she was interested. I assured them that we would attempt to structure an offer that "they couldn't refuse."

Our consultant helped put together an offer that would allow each of the five people to purchase 5 percent of the firm's stock in a private transaction from my partner and myself. The goal was to structure a deal that required each of them to stretch financially, put their "skin in the game," but still make it affordable. The offer was structured for each of them to make an initial down payment and pay out the remainder in a three-year note at a very favorable interest rate.

Presently, we are three years into our ESOP with three new stock owners (we lost two people along the way), each with 5 percent stock, and I feel we have ensured a bright future for the firm, clients, and staff. As mentioned at the beginning, each firm is different and what has worked so far for our firm may or may not work for you, so seek help from a specialist and structure a transition program that works for you.

Now let's examine the options and details of ownership transition.

PLANNING FOR OWNERSHIP TRANSITION—EXPLORING THE OPTIONS

Planning for ownership transition includes considering whether you will sell internally to staff or externally to another firm. There also is the "do nothing option," which essentially leads to liquidation and closing the doors. Here are several considerations owners have to weigh in deciding whether to sell to the staff, sell to outsiders, or to just "close the doors."

Considerations for Selecting the Method of Sale

- *What is the time frame for the existing owner(s) to sell out?* How old are they and what is their time frame? Is it 5–10 years or have they decided that they want out as soon as possible? Obviously, a transition that has a longer time frame allows the current owners more time to plan and evaluate more options. It also allows for several false attempts before the owner(s) are faced with accepting a short-term option that may not be beneficial to them, their clients, or the firm. In our case, we started the process about 10 years ago, worked at several attempts over that period, and now appear to have a transition plan in place that will take about five more years. Owners who wait too long may be faced with having to sell without a plan which serves them well.
- *Is the staff interested and are there sufficient leaders?* Do you have experienced staff who can take over management and leadership with training and mentoring? If you do, are they willing to put in the time and money to do it or would they rather work for someone else? In other words, do they have an entrepreneurial spirit or are they risk averse?
- *What about company culture change?* Does your company have a unique culture that may be destroyed if the firm is sold externally? Our firm does, and culture was one of the main motivators for us to seek an internal sale. Our culture includes: flex time, paid health insurance, continuing education reimbursement, paid professional society membership dues, Friday afternoons off between Memorial Day and Labor Day, and a company ski team which races every Wednesday from January through March. The relatively small size of our firm, about 20 people, also provides a great family atmosphere. Each of us knows other staff's spouse or significant other as we'll as their children. I believe most of this would be lost if we were to sell to a larger firm.
- *What are the tax implications for buyer & seller?* While this consideration may not be a deal maker or breaker, it can have significant impact on the sale. Sale of stock to a qualified ESOP allows the capital gains tax, typically paid by the seller, to be deferred as long as the stock is

reinvested in stock of domestic corporations. Private sale of stock may require that taxes be paid several times if the stock is purchased with employee bonuses. First, the employee pays taxes on their bonus as regular income, and then the owner who sells the stock pays capital gains tax on the proceeds received. The tax implications of any sale should be reviewed with your tax adviser.

- *What are the guarantees that the seller will be paid?* Once the seller closes the transaction, he needs guarantees that he will be paid. The transaction usually includes a cash down payment and a note for the outstanding balance. As long as the firm continues to be profitable, payments to the seller are usually made in a timely manner. It is when there is an economic downturn and cash flow becomes tight that the seller may experience problems. The seller should seek assurance from the buyers that his note will be the first invoice paid each month. If possible, the note should be paid off during the transition period, while the seller is still actively involved in the firm. The most disappointing thing that can happen to a seller is to have to come out of retirement and return to the firm in order to protect his remaining assets.
- *What about the unforeseen, death or disability?* Nobody likes to consider this but, often senior owners have health issues or sometimes a death occurs before a senior member of the firm retires. The plan for these occurrences needs to be spelled out in detail in the stockholders' agreement. Usually, it includes the firm holding a life insurance policy on key senior staff that would pay the firm an amount large enough to purchase the senior owner's stock should death or disability occur.
- *How will the deal be financed?* As previously mentioned, the transaction usually includes an initial down payment with the balance secured by a note. The note may include a bank loan at favorable interest rates secured by the firm, which agrees to buy back the stock and pay off the loan if there is a default. Financing also may include a private note between the purchaser and seller of stock at a favorable interest rate and repayment over a reasonable time period. Other financing methods such a deferred payments and exchange of assets will be discussed later in this chapter.
- *How are the clients affected?* The goal of any ownership transition should be to have minimum impact on the firm's clients. If the decision is made to transition internally, the departing owner(s) needs to introduce their clients to the new owners over a sufficient period of time for the clients to get to know them and to develop a comfortable relationship with them before the senior owner(s) leaves. The departing owner(s) needs to assure their clients that the new owners will provide the same quality and level of client service that previously was furnished. The

departing owner(s) also needs to train the new owners so that they understand the type of special treatment the departing owners may have given to a particular client.

- *How is the rest of the staff affected?* If an internal transition is taking place, the goal is to minimize disruption of the staff if the transition is going to be successful. The staff needs to be kept abreast of the transition planning and their doubts and fears dealt with whenever possible. They, too, need to have confidence in the new owners. In each of our transition attempts, we lost some staff for a variety of reasons. Some may have felt they were being forced to consider ownership when they really only wanted to be just a staff engineer or surveyor and others had their own reasons, but it appears, at least in our experience, some staff turnover may be the result of any transition attempt.

If external sale or merger is the plan, it is usually kept secret until the transaction is done. This may have an impact that causes some staff to feel that they have been betrayed and they must move on.

Answers to each of the following questions will help you determine whether or not you have the possibility of making a successful internal ownership transition.

- Internal transition—some initial questions to answer:
  - Growth or no growth?
  - Can we affect management as well as ownership transition?
  - Have potential owners been identified?
  - What is the schedule for the transition?
  - How will the transaction be financed?
  - What criteria has been established to qualify new owners?
  - Do we have a leadership development program?
  - Will the clients be well served?
  - Will the staff be supportive and confident in the new owners?

If you have good answers to the preceding questions, proceed with the following additional considerations.

- Internal transition—some important financial considerations:
  - Does the firm have successful history of growth and generating profits?
  - Do you have a recent valuation of the firm?
  - What financial options may be considered by the stock seller(s)?
    - Employee Stock Ownership Plan (ESOP)

- Private sale of stock to key staff
- Bank loan versus private notes
- Bonus stock
- Stock options
- Stock redemptions
- Deferred compensation
- Annuities
- Exchange of property

Once the financing options have been explored and the preferred method selected, proceed to planning the firm's future. Keep in mind that you may be asking new owners to plan 10 to 15 years into the future, which may be one third to half of their life. Planning takes the experience and historic perspective from senior leaders to help new owners form the vision for the future.

Internal Transition—Planning for the Future

- Establish the planning committee.
- Review the firm's mission statement. Has it changed?
- Seek input from the new owners to develop the vision of the future.
- Determine the firm's strategic objectives.
- Start with a 1–2 year operating plan.
- Begin strategic/market planning.
- Establish strategic and operational action plans.
- Review and update the plans regularly.
- Develop a management transition plan.
- Develop a stock transfer agreement.

If, after reviewing the preceding criteria for internal ownership transition, you are not convinced that it will work in your firm, you should consider external sale of the firm. Warning: investigate this option very carefully. What sounds like a good deal, upon further investigation may not be so. Seek professional help, including an independent determination of the value of your firm and legal help to guide you through the process. While an external sale or merger may sound like the easiest and quickest way to retirement for the seller, be aware that many of these transactions ultimately do not work.

- External transition—options
  - Merge with a firm of similar size, interest, and philosophy.

- Sell to a larger firm that wants your clients, staff, market niche, geographic location, and so forth.
- Sell to a large company not in the engineering or surveying industry.
- External transition—advantages
  - Departing owner(s) may be able to leave sooner.
  - The total price for the firm may be greater than an internal sale.
  - The continuity of operation is maintained.
  - Cash flow problems may be reduced or eliminated.
  - Needed staff can be obtained.
  - Needed technical expertise can be obtained.
  - Planning and marketing expertise may be obtained.
  - New markets may be introduced.
  - There is access to larger projects.
  - Access to a broader geographic area.
  - Economic fluctuations may be moderated.
- External transition—disadvantages
  - A sale or merger creates uncertainty with the staff.
  - Key staff may leave the firm.
  - Family-oriented culture and commitment may be lost.
  - Administrative staff may be reduced or eliminated.
  - The new firm may not have the same commitment to the community.
  - Payout to seller(s) may be done with future profits.
  - Bonuses may be reduced.
  - Profits may not be reinvested in the firm.
  - Seller(s) may loose their job or be tied to an employment agreement.
  - The change can create problems with clients.
  - Benefit programs may be consolidated and not as generous.
  - The different culture needs to be understood.
  - Different philosophies may create integration problems.
  - Overall consolidation may result in closing the local office.
- The final option—when no choice is left, close the doors.
  - "Fire sale"—Liquidation
    - Usually occurs when the owner(s) are in poor health, or through death or disability.
    - May be forced by the estate of a majority owner.
    - May occur when the owner has waited too long to sell and clients have dwindled, and the business has declined and is no longer viable.

- May happen when there are no buyers internal or external for whatever reason.
- It may take longer and be more difficult than you planned.
- What is the value at a "fire sale"?
  - Usually the clientele, equipment, and records.
  - Client value is usually less than expected.
  - Some equipment may be able to be sold at auction.
  - Records may have very limited value.
  - There may be some work-in-process that can quickly be finished and invoiced.
  - Accounts receivables.
  - Remember the firms debts must be paid off, too.

See Table 10.1 for an overview of the advantages and disadvantages of different methods of selling a firm.

- The nitty-gritty details of internal transition—the author's preferred method
  - Answering the initial questions

As mentioned previously, the internal ownership transition is my preferred option for a variety of reasons, not the least of which is that it allows me to stay involved in the firm on my own terms, contribute in a productive way, and participate rigorously in mentoring and development of the next generation of leaders and managers, which I thoroughly enjoy. Previously, I outlined some initial questions that must be answered by owner(s) of a firm seeking to sell internally.

Growth In my opinion, "Growth or no growth?" is the most important question to be answered in deciding if internal ownership transition will work for your firm. If fact, one of our failed transition attempts included a candidate for ownership who felt the firm should not grow. Needless to say, the other potential owners felt differently.

New owners generally feel that growth of the firm is important. Most of the time they are young and see that the only way to produce higher salaries and profits in the future is to grow the firm. I believe it also is necessary to continue to challenge younger members of the staff and to provide an upward career path for them. Growth also can include bringing in larger, more challenging and more complex projects to the firm as well as increasing staff and income.

However, growth can't be a dream or a figment of one's imagination. The firm must have historic growth and profits in order to convince the new

Table 10.1 Advantages & Disadvantages of Different Methods of Sale

Transition Type	Advantages	Disadvantages
Internal Transfer	Maintains continuity	Second generation lacks funds for buyout.
	Improves moral & loyalty	Owners may have to accept an extended payout.
	Employees see path for advancement	Second generation must be convinced that buying stock is good.
	Employees enjoy prestige of ownership	Buyout is accomplished with after-tax dollars.
	Young employees likely to remain with the firm	Second generation may have to incur debt to finance purchase.
	Easier to retain current clients	Second generation may perceive that it is easier to start their own firm.
	Principals control timing	
External Transfer		
	Original owners may have options of immediate or deferred payment.	Leads to uncertainty in staff and possible defections
	New owners may inject additional funds into firm for expansion.	New owner's philosophy of operation may alienate some present clients.
	New owners may bring new management techniques and make improvements in the organization.	Competitors may take advantage of change to spread rumors about the firm.
	New owners may have wider contacts and improve marketing efforts.	New owner without experience in managing a professional firm needs to learn and adapt
		Usually involves the immediate loss of control by former management
Liquidation		
	Instant cash for sale of furniture and equipment	Total sale value doesn't include all assets
	Probably the shortest time frame for sale	Clients may be left without continuity of services.
		Value may be substantially lower than other methods
		Staff is out of a job immediately.

Adapted from Exhibits A & B – *Ownership Transfer Options & Strategies* by Paul M. Lurie & Lowell V. Getz. Birnberg & Associates, 1994 owners that continued growth is possible. Most firm valuations consider past growth and project it into the future as part of determining the firm's value. The key for internal transition is for the departing owner(s) to train and mentor the new owners to sustain the growth pattern of the past. This can be done through continued effective financial management and strategic planning for the future.

Leadership and the Management Transition If you have all followers and no leaders internal ownership transition is going to be very difficult. I thought I might have this problem in our firm, so this is how we went about determining if we had future leaders and managers. I knew if I could determine that we had leaders, I could develop them as managers and, in fact, some already were serving as mangers. I started out with no preconceived notion as to who the future leaders might be. The younger members of our staff who presently were serving as middle managers obviously had shown potential earlier in their career. Would they step up and show that they were entrepreneurs and risk takers and would be able to lead the firm into the future?

As I previously mentioned, over the period of a year, I held leadership seminars once a month after work in order to determine who was willing to put I the extra effort. This was my first evaluation criteria since I've found that firm leaders need to be willing to put in more effort and sometimes extra hours in order to be successful. I also was looking for people who said "we" rather than "you" and "our firm" rather than "your firm" during discussions. This simple demonstration of mindset showed me that people were thinking of the bigger picture and embraced a team approach. The first few seminars were attended by over half of the members of the firm but soon attendance dropped to 5–8 people who consistently showed up and contributed positively to each seminar. From the group we identified five people with potential to whom we offered ownership in the firm. The others, while very interested and mature beyond their experience, needed more technical experience and licensing. If their enthusiasm continues, they will be the core of the next generation of new owners.

Potential Owners Identified The five people, who after a year were offered ownership, also met certain minimum criteria that had been established many years ago as part of earlier attempts. These are shown in the following text box.

We and our ownership management consultant began to work with this group to help them understand what the requirements would be for their "skin in the game."

Minimum Requirements for New Owners

1. Experience with the firm – Minimum of five years of increasing level of responsibility.
2. Licensure – Obtained professional license when applicable and maintains license in good standing.
3. Technical competence – Has a high level of technical competence that is recognized by peers.
4. Management skills – Has proven management skill as a project manager or department leader. Proven ability to bring projects in on time and within budget. Also, has proven client service skills. Goes the extra mile. Is a problem solver.
5. Marketing capability – Has proven ability at maintenance marketing, prospecting, lead development, and successfully writing proposals and closing the deal. Has proven marketing presentation skills.
6. Professional involvement – Participates in professional societies at a high level, doesn't just attend meetings.
7. Community involvement – Active in community affairs and gives back to the community.
8. Education – Minimum of bachelor-level degree in engineering or surveying. Higher-level degree desirable. MBA a definite plus.

Any of the preceding criteria may be waived in special circumstances.

After the new potential owners were identified, they began to participate in meetings with our consultant to better understand how the value of the firm was developed and in crafting and reviewing the offer of stock letter of intent and the method of payment.

Financing the Initial Transaction The initial stock offering to the potential new owners was only 15 percent of the company stock, 5 percent to each new owner. We sold 3 percent of the stock at a discount to its calculated value, and we gifted the final 2 percent to them. After the initial down payment, the remainder was financed through a personal note for 3 years at a very low interest rate. This was a very attractive offer designed to allow the new owners to buy into the firm at a very affordable price. Future offers for larger amounts of stock may necessitate a bank loan.

Schedule Following acceptance of the offer and the new owners signing a letter of intent, we began working on the stock agreement. The goal was to cover all of the necessary items but keep the agreement simple and complete the transfer within 6 months. We felt that a very important milestone had been reached in the history of our firm and held a small celebration to introduce the new owners to the rest of the firm's staff. This will be followed up with the appropriate press releases and announcements to our clients.

New Owner Training Once the initial transaction took place, we started a series of regular meetings to introduce the new owners to the intricacies of day-to-day operations of the firm. We started with a description of our type of corporation and the bylaws, described how we pay taxes, worked our way through how the annual budget is developed, and discussed how the bonus system works. The new owners are now members of the board of directors and will attend regular directors meetings. I envision continuing the new owners meetings and training through the length of my transition period.

Client Service In each case, clients are currently working with our new owners who also serve as project managers. Now that the new owners have been identified, we will introduce them as such, to our clients in order to assure them that they will be served with the same high level of service that they are presently receiving. One of the challenges is to get clients to recognize that the new owners have the motivation, authority, and ability to serve them. It seems that clients work well with them up to a certain level of problem solving or issue resolution, and after that they want to deal with an owner. In order to earn their respect, the new owners are being mentored to increase their level of experience in solving the difficult problems. Respect must be earned; it can't be mandated.

Staff Support I believe that we conducted the initial ownership transition process with the full knowledge of and disclosure to the rest of our staff. While they were not involved in the valuation of the firm discussions, the details of the offer, or the transaction itself, we have been very willing to discuss the process with anyone who asks. I believe that the staff has complete confidence in the new owners and will be supportive of their actions. After all, we need to keep them motivated in order to develop the next group of stock purchasers.

Financing the Deal—Important Financial Considerations

It almost goes without saying that our profession doesn't pay top salaries. This is so for a variety of reasons, but the most important things is that because of their salaries, most potential new owners don't have a lot of money available to by stock. This creates the need for some alternative methods of financing if the transaction is going to be successful. I'll discuss some of the more viable options for small engineering and surveying firms.

Employee Stock Ownership Plan (ESOP) Given our experience with earlier failed attempt at transitioning ownership to the next generation, my partner and I decided to pursue an ESOP. Since the past failed attempts each involved potential owners not accepting our offer of stock for a variety of reasons, we decided that an ESOP was needed in order to get everyone in the firm initially involved in ownership. After all, we believed that after our departure, it would truly be "their firm." An ESOP is a trust in which the profits of the firm are contributed to the trust and used to purchase stock (from the seller) in the name of the individuals who participate in the program. As long as the IRS rules are followed (a minimum of 30 percent of the firm must be owned by the ESOP), the plan is a qualified employee benefit plan. It allows the seller to defer tax on the capital gain from the sale as long as the proceeds are invested in stock of a domestic corporation. Typically, this means stock of solid blue chip corporations, since the stock is usually used as collateral for the ESOP loan. The ESOP trust is administered by a trustee who is appointed by the firm's board of directors. The trustee receives the profit contributions and is responsible for seeing that they are used to pay off the loan and allocate stock, according to the approved formula, to each individual's account.

An ESOP can be either leveraged or nonleveraged. A leveraged ESOP is financed by an outside bank loan while a nonleveraged one is internally financed. Since most small firms don't hold a lot of excess cash, a leveraged ESOP is the typical way that the transfer is financed. The IRS rules allow the firm to deduct both principal and interest as well as providing certain advantages to the lending bank. This normally results in a favorable interest rate for the loan. ESOPs are very complicated, so be sure to consult an expert when setting one up.

Private Sale of Stock A private sale of stock is carried out between a seller and a buyer. Usually, this is a retiring owner who is selling to a new owner. It doesn't include selling stock back to the firm's treasury or acquiring stock from the firm. It may involve a private note, at a very favorable interest rate, as previously mentioned, for the initial sale of stock to new owners, or it may involve a bank loan. If a bank loan is used, the stock is used as collateral. Usually, the firm agrees to guarantee that they will repurchase the stock in case of default. The new owner assumes full responsibility for the loan unless he/she defaults. All stock sales and purchases are governed by the terms of the stock sale agreement. Another method of generating cash to purchase stock is for the buyer to borrow from their 401(k) account.

I don't recommend this, since it has the potential of reducing the new owner's retirement benefits and incurs a penalty if there is a default.

Bonus Stock Bonus stock is issued, from the firm's treasury, instead of cash bonuses. This is an easy way of allowing key employees to acquire the firm's stock. It may include redistributing stock redeemed by the firm, which is buying back a retiring owner's stock, or it may be a new stock issue. Bonus stock is taxed as ordinary income at its market value at the time of distribution. An advantage is that it conserves cash and eliminates the added transaction of having the employee purchase stock with their bonus. It may appear to have less value than a cash bonus and therefore not be desirable to some staff.

Stock Options These are also known as "incentive stock options." They allow employees to purchase stock in the future at the market value established at the time the option was granted. If the firm is growing, the employee realizes the benefit of the increased value of the stock over the price at the time the option was given. If the firm is growing consistently, this can be an attractive benefit. Options are generally used to attract senior high-salaried employees and to retain key staff members. They are not very useful for internal ownership transition, since the receiver of the option still has to exercise it, and many young people don't have the means to purchase the stock even at the option price.

Stock Redemptions Special IRS rules apply to stock redemptions so that the redemption doesn't appear to be a bailout or a means to pay only capital gains tax rates on a dividend. In order to qualify as a redemption, the seller has to completely terminate his/her interest in the firm or have a substantial disproportionate reduction of interest in the firm. This is defined as:

1. Immediately after the redemption the shareholder owns less than 50 percent of the voting stock and,
2. He owns less than 80 percent of what he owned prior to the redemption.

The redeemed stock remains as treasury stock until it is issued to new shareholders. Redemptions require excess cash in the firm treasury; otherwise, a bank loan must be taken to redeem the stock of a retiring owner. Redemptions may involve installment payouts. In either case, it may be seen as a burden on the firm's cash flow with little benefit to the new owners.

When an owner is retiring without new owners being added, the firm may buy the stock at full value and return it to the treasury, in a single transaction, for later distribution or the remaining owners may purchase the retiring owners stock, in multiple transactions, in proportion to their current interest. In either case, the retiring owner receives full value for his/her stock, which is then taxed as a capital gain.

Deferred Compensation A retiring owner can sell stock to new owners at book value and then be compensated for the remaining value through a deferred compensation plan after retirement. In order to protect the plan, funds are withdrawn from the firm and placed in a trust, which will pay out the seller over a period of time. The trust is owned by the firm and may be subject to claims by creditors. The firm does not get an expense deduction for contributions to the trust, and the new owners are not taxed until the stock is released from the trust. This type of plan could be beneficial to the new owners if the value of the stock increases in value over the deferred payout period. It appears to have little value to the stock seller other than being a method of obtaining some additional compensation when other options do not appear possible.

Annuities This is where the firm buys an insurance policy on the life of the major stockholder. The value of the annuity rises each year as it paid off. Once the annuity reaches the value of the stock, it is exchanged for the retiring owner's stock. The retiring owner then cashes in the annuity for a continuing stream of payments over a period of time.

Exchange of Assets Where a firm owns its own office building and exchange of assets may be possible. The firm deeds the building to the retiring owner in exchange for equal value of the owner's stock. This is a taxable transaction for both parties who pay capital gains tax on the difference between the selling price and the basis for the building or the owner's stock. The retiring owner's stock then may become treasury stock for redistribution. This type of transaction is not common, since most firms do not own their real estate and even if they do the value of the real estate needs to be close to that of the stock being acquired. This can be a complicated transaction and legal as well as tax advise is necessary.

Firm Valuation

Before the ownership transition process can progress to a specific offer, the value of the firm needs to be determined. Some firms sell stock to new owners at the "book value" shown on their balance sheet. While this is a very simple way to carry out the transaction, it doesn't adequately reflect the true value of the firm or adequately compensate the retiring owner. The firm's books may be kept on a cash basis and items such as work-in-process, accounts receivable, good will, potential profits for ongoing work, and projected income for the future are not considered in book value. In addition, if ownership transition is going to include an IRS "qualified plan," the valuation must be done by an independent third party. The formula or method of determining firm value needs to be specified in the buy/sell stockholders agreement.

What is the fair value of an engineering or surveying firm? Since most small firms are privately held, the value is much more difficult to determine than that of a public company whose stock is traded each day on the New York or other stock exchanges. Intuitively, we know that fair market value is what a willing buyer will pay and a willing seller will sell for in an "arm's-length" transaction. There are eight common factors which, are often considered when valuing a private company:

1. *The nature of the business and its history since inception.* Is the firm a growing enterprise or has it been declining recently? How long has the firm been in business?
2. *The overall economic outlook in general and for the specific industry.* Is the general economy doing well or is it in a recession? Does the specific industry follow the general economy or is it countercyclical?
3. *The book value of the stock.* What is the current financial condition of the business? Are there large accounts receivable and payable? How is work-in-process valued?
4. *The earning capacity of the business.* Does future growth project into profit growth?
5. *The dividend paying capacity of the company.* In the case of private companies, have bonuses been paid regularly?
6. *Goodwill or intangible value.* Is the reputation tied to one person?
7. *Size of the block of stock to be valued.* What is the reason for valuing the company?
8. *The market price of the stock of companies of similar size and business.* How can this be determined?

Other factors to be considered are the date of valuation, backlog of work, potential liability or claims, competition, and vulnerability of the industry overall. Intangibles such as goodwill, reputation, years in business, geographic location, current management and staff, type of work performed, number of repeat clients, market niche specialization, and investment in technology also play a role in the firm's value.

Since it is so difficult to determine an exact value of a small privately held firm, it only makes sense to look at a variety of methods and compare the different values derived from each. What are some of the different methods?

- *The market value approach.* Since there is no public market for small engineering and surveying firms, this is where an experienced valuation consultant comes in. If you hire a consultant who does many valuations of similar firms each year, he/she knows how much similar firms have sold for. The consultant will apply the preceding factors and make adjustments that are unique to your firm, and based on his experience will come up with a value very close to what the firm would sell for in a similar transaction. The market value approach reflects the value a willing seller will receive and a willing buyer will pay, assuming that both are equally informed and neither is under any obligation to buy or sell the firm.

  The market approach actually includes two approaches. First, it includes research of publicly traded engineering companies (there are no known publicly traded surveying firms) to identify potential factors or guidelines from which to draw an "inference of value." Second, if the consultant is experienced in the valuation of similar firms, and mergers and acquisitions of small private firms, he will search for comparable transactions to analyze certain information "deemed comparable." In publicly traded companies, comparables may include price to earnings ratio (P/E), the price paid per share of stock divided by its earnings per share; price to book value ration (P/BV), the price paid for the firm divided by the book value from the adjusted balance sheet; earnings before interest and taxes (EBIT); and equity to debt ratio (E/D). Under the mergers and acquisitions approach, actual transactions are used to compare similar ratios to those in public companies.
- *Income-based approach.* This approach utilizes the capitalization of historic earnings or cash flow and the discounted future cash flow of future earnings. Future earnings are discounted to the present-day value by using a reasonable discount rate (market-based rate of return) and the expected growth rate of the firm. Earnings adjustments may include historic volatility; claims or professional liability history; amount of work obtained on a qualifications, rather than fee bid basis; number of proposals written and hit rate; and overdependency on a few clients. The normalized earnings are used to determine the earnings-based value of the firm. Expectations for growth are not necessarily used in determining the cash flow stream.
- *Asset-based approach.* This is a commonly used method for valuing small firms when liquidation is eminent. It considers the value of the

firm's hard or tangible assets, including cash on hand, factored accounts receivable, undepreciated or market value of vehicle and equipment, and value of the firm's real estate, if any. It starts with the firm's "book value," which is defined as assets minus liabilities. Little value is added for the company's labor, work-in-process, prospective contracts, or future earnings. Various adjustments are made, including removing real estate value, if it is included, adding back excess salaries and bonuses, removing uncollectible receivables, unbillable work-in-process, and excess contributions to 401(k) or other profit-sharing plans. The end result is a firm valuation based on adjusted book value.

The adjusted net asset value differs from the market and income approach in two ways. First, it focuses on the value of individual assets and liabilities. Second, it is normally only used to value a firm that is considering liquidation.

The methods just discussed are the most common ones used to establish the value of privately held engineering and surveying firms with the most frequently used being the market approach and the income approach. In the market approach, more weight is usually given to comparisons developed in the merger and acquisition basis, since it more closely reflects small privately held firms. If the proper adjustments are applied the firm value from both, the calculations should be very close. Often the average of the two values is used as the stated value of the firm (see Table 10.2).

The Offer and Agreement

Following the firm valuation and the identification of perspective new owners, an offer of stock is made, it is accepted, and the terms of the agreement are worked out. Outlined here are the characteristics and samples of the key documents that need to be completed in order to consummate the transfer.

Table 10.2 Methods of Valuation

Valuation Method	How used
Market Value Approach	Not easy to apply to privately held firms. Adjustments needed. Publicly owned firms for comparison only.
Income-Based Approach	Uses historic earnings and discounted cash flow of future earnings. Most common method used for private firms.
Asset-Based Approach	Used to determine liquidation value of vehicles and office equipment. Generally used after other methods fail and owner has no choice.

The Stock Offering Memo

ACME Engineers & Surveyors, Inc.

Letter of Intent/Term Sheet

Offering:	Acquisition of the common, voting stock of ACME Engineers & Surveyors, Inc.
Seller:	Mr. John Q. Good and Mr. Joseph R. Traverse.
Purchasers:	Mr. James Parker, Mr. Samuel Fine, and Ms. Sharon Showers
Number of Shares:	Up to 30,000 common shares in aggregate are offered, representing 15.0% of the equity of ACME Engineers & Surveyors, Inc. after the transaction. James Parker is offered a 5.0% interest in the Company, or 10,000 shares. Mr. Parker shall purchase the initial 3.0%, or 6,000 shares based on the share price below with the remaining 2.0%, or 4,000 shares, to be gifted to him. Samuel Fine is offered a 5.0% interest in the Company, or 10,000 shares. Mr. Fine shall purchase the initial 3.0%, or 6,000 shares based on the share price below with the remaining 2.0%, or 4,000 shares, to be gifted to him. Sharon Showers is offered a 5.0% interest in the Company, or 10,000 shares. Ms. Showers shall purchase the initial 3.0%, or 6,000 shares based on the share price below with the remaining 2.0%, or 4,000 shares, to be gifted to her.
Share Price:	$X.XX per share—the approximate result of 80% of the valuation as of December 31, 2007. This share price shall apply to the first aggregate 9% that is being acquired by the purchasers (this represents the 3% being acquired by each of the purchasers by December 31, 2008). The share price for the remaining aggregate 6% (those shares being gifted or bonused in 1% increments) shall be determined when those shares are transferred from sellers.
Purchase Amounts:	James Parker's purchase amount is approximately $ XX,XXX. Samuel Fine's purchase amount is approximately $ XX,XXX. Sharon Showers's purchase amount is approximately $ XX,XXX.
Payment Terms:	Purchasers will purchase offered shares with 20% cash ($X,XXX) at closing and finance the remaining amounts with a note financed by the sellers. The terms of the note will be 3.0% interest per annum over a two-year period.

Repayment:	Purchasers shall repay debt with quarterly payroll deduction and use of potential incentive compensation.
Actions Precedent:	Company and shareholders shall take all corporate, shareholder, and board actions required to ensure that stock offering is in compliance with all laws and regulations guiding such transactions.
Documents:	Purchasers will execute customary documents, including: Stock Purchase Agreement, Shareholders Agreement, Shareholder Indemnification Agreement, and other documents customary for such stock purchase.
Closing:	Closing shall take place on or before December 31, 2008.

Our signatures below represent our letter of intent to follow through with this transaction. We understand that as the legal documents are prepared, further discussions may occur, specifically with respect to how shares are valued in future years.

_______________ _______________

Seller Date Witness

_______________ _______________

Purchaser Date Witness

Stock Purchase Agreement or Letter of Intent (LOI) The Shareholder Agreement The deal has been consummated, and now all that is left are the nitty-gritty details. Don't ignore them or gloss over them. The shareholder agreement is the final document, which concludes months or even years of effort by owner(s) to transition the firm to the next generation, but it needs to be detailed in order to address all of the "what ifs." Remember that all parties to the agreement, sellers and buyers, must agree to be bound by the terms of the agreement.

Some of the details that need to be addressed in the agreement are as follows:

1. *Basic definitions.* The terms discussed throughout the agreement should be clearly defined up front so that everyone who reads it knows what they mean. Important items to be defined include:
   (a) Book value of the company
   (b) Buyer
   (c) Cash flow
   (d) Closing date

(e) Competes with the company
(f) Death of an owner
(g) Departure of a shareholder
(h) Involuntary departure
(i) Permitted transfers
(j) Purchase price
(k) Retirement
(l) Sale or transfer
(m) Seller
(n) Stock
(o) Termination
(p) Total disability
(q) Triggering event
(r) Value of the company (firm)
(s) Voluntary departure

2. *Stock.* What happens to the stock also needs to be clearly discussed.
   (a) How many shares of stock are there?
   (b) Will more stock be created as part of the transaction?
   (c) Are there different classes of stock?
   (d) How is stock voted?
   (e) Are there restrictions on transfer of the stock?
   (f) How are stock certificates held and what is the text for a stock certificate that meets the legal requirements?
3. *Insurance policies.* The transaction should address whether or not life insurance policies will be taken out on key stockholders. This is usually done to guarantee that the firm will have enough money (and not create a financial burden on new owners) to purchase the stock of a major stockholder should there be a death before the entire transition is complete. The beneficiary is usually the corporation who also pays the annual premium.
4. *Voluntary transfer of stock.* What constitutes a voluntary transfer and who has the right of first refusal? How long is the offer period? Does the company always have the right of first refusal? Can a departing owner sell his/her stock to the company ESOP without offering it to other owners first?
5. *Obligation to sell.* There are several key triggers that may create obligations to sell stock. This usually happens when a stockholder reaches retirement age, generally between 60–65. Often the stockholder is

required to begin to sell stock at the triggering age whether or not they actually retire. Death or total disability also may trigger an obligation to sell stock back to the firm. The time frames for repurchase of the stock also should be defined here. Generally, the time frame is set up to not create financial difficulty for the firm.

6. *Purchase price of the stock.* The method of valuing the company and the formula used need to be part of the agreement. Determination of the sale price of stock and any discounts applied also need to be specifically stated here.
7. *Closing date.* The original closing date is not the only one that needs to be discussed. Closing date after a death, disability, voluntary transfer, or termination also need to be specified.
8. *Provisions for purchasing shares and terms of payment.* These also need to be included as well as limitations on pledging shares or using them as security.
9. *Control and management.* The composition and power of the board of directors needs to be clearly defined as well as the fact that share ownership doesn't necessarily include a position in upper management. It also needs to be clear that the stockholder agreement is not an employment agreement and a stockholder's job is not guaranteed.
10. *Confidentiality and noncompete obligations.* It almost goes without saying that new stockholders must agree to not leave and start a direct competitive firm. Owners also must agree to protect company trade secrets, proprietary rights, and other confidential firm information. Sanctions for violations of this clause also may be considered.
11. *Miscellaneous items.* Finally, the agreement should include items that discuss the enforceability of the agreement, the state of governing law, a severability clause, and provisions for assignment, superseding, or termination of the agreement.
12. *Signatures.* The final page of the agreement should be the signature page for sellers and buyers. Check the state laws regarding the requirements for witnessing signatures and notarizing them.

So there it is, ownership transition, including all of the options. This is a very new field for many retiring owners. Many of us are familiar with past generations who just slowed down their work load and liquidated their practice upon retirement. I hope ownership transition will be a rewarding experience for all involved, including clients and firm staff as well as the departing owner(s) and new stock purchasers.

BIBLIOGRAPHY

American Consulting Engineers Council, *Financial Management and Project Control for Consulting Engineers*, Washington, DC: CPA, American Consulting Engineers Council, circa 1980.

Bachner, John Philip, *Practice Management for Design Professionals, A Practical Guide to Avoiding Liability and Enhancing Profitability*, Hoboken, NJ: John Wiley & Sons, Inc., 1991.

Bergeron, H. Edmund, "From Engineer to Entrepreneur: Making the Transition," *ASCE Journal of Management in Engineering*, March/April, 1998.

Berger, Louis, " Emerging Role of Management in Civil Engineering," *ASCE Journal of Management in Engineering*, July/August, 1996.

Black, Henry Campbell, *Black's Law Dictionary*, fourth edition, St. Paul, Minnesota: M.A.West Publishing Co., 1968.

Boone, Louis E. and Kurtz, David L., *Management*, fourth edition, New York: McGraw-Hill, Inc., 1992.

Council of American Structural Engineers, *Business Practice and Risk Management Newsletter*, August 29, 2008.

Covey, Stephen, *The 7 Habits of Highly Effective People*, New York: Free Press, 1989.

Charan, Ram; Drotter, Stephen; and Noel, James, *The Leadership Pipeline: How to Build the Leadership Powered Company*, Hoboken, NJ: Jossey-Bass/John Wiley & Sons, 2001.

Dixon, Sheila A. (ed.), *Lesson in Professional Liability: A Loss Prevention Handbook for Design Professionals*. Monterey, CA: XL Insurance – Design Professional Monterey, 2004.

DPIC Management Services Corporation, *Successful Internal Ownership Transition*, Seminar handout notes, Chicago, IL, May 6, 1999.

Fields, Edward, *The Essentials of Finance and Accounting for Nonfinancial Managers*, New York: American Management Association, 2002.

Frost, Susan E., *Blueprint for Marketing: A Comprehensive Marketing Guide for Design Professionals*, second edition, Portland, OR: SEF Publications, 1995.

Getz, Lowell and Lurie, Paul M., *Ownership Transfer Options and Strategies*, Washington, DC: American Consulting Engineers Council, undated.

Goleman, Daniel, *Emotional Intelligence*, New York: Bantam Books, 1995.

Harding, Ford, *Creating Rainmakers*, Avon, MA: Avon Media Corporation, 1998.

Katz, Robert, *Skills of an Effective Administrator*, Boston, MA: Harvard Business Review, September 1974.

Lewis, James P., *Project Planning, Scheduling, & Control*, third edition, New York: McGraw-Hill, Inc., 2001.

Litka, Michael P. and Inman, James E. *The Legal Environment of Business*, third edition, Hoboken, NJ: John Wiley & Sons, Inc., 1983.

McGhee, Sally, *Take Back Your Life*, Seattle: Microsoft Press Learning Center, 2003.

Mitchell, Terence R., *People in Organizations: An Introduction to Organizational Behavior*, second edition, New York: McGraw-Hill, Inc., 1982.

Phillips, Barbara Ashley, *Finding Common Ground: A Field Guide to Mediation*, Halfway, OR: Hells Canyon Publishing, 1994.

Pruitt, John (executive summary), *2006 Operating Statistics Survey for Project Focuses Professional Service Firms*, Herndon, Virginia: Deltek Systems Inc., 2007.

Rachlin, Robert and Sweeny, Allen, *Accounting and Financial Fundamentals for Nonfinancial Exectuives*, second edition, New York: American Management Association, 1996.

Safford, Dan, *Proposals: On Target, On Time*, Washington, DC: American Consulting Engineers Council ACEC Press, 1997.

Stit, Fred A. (ed.), *Design Office Management Handbook*, Arts & Architecture Press, 1986.

Stone, David A. *Marketing in the 21st Century for Design Professionals*, Washington, DC: ACEC Press, 2002.

Sweet, Justin, *Legal Aspects of Architecture*, third edition, St. Paul, Minnesota: West Publishing Co., 1985.

Robert Tannenbaum and Warren Schmidt, " How to Choose a Leadership Pattern," *Harvard Business Review*, May-June 1973.

Thomas, Brian Scanlon, *Marketing of Engineering Services*, London: Telford, Ltd., 1988.

Walesh, Stuart G., *Engineering Your Future, The Non-Technical Side of Professional Practice in Engineering and other Technical Fields*, second edition, Washington, DC: ASCE Press, 2000.

Weingardt, Richard, *Forks in The Road*, Denver: Palamar Publishing, 1998.

INDEX